COMMUNICATING
SCIENCE
A Practical Guide for
Engineers and Physical Scientists

COMMUNICATING SCIENCE
A Practical Guide for Engineers and Physical Scientists

Raymond Boxman
Tel Aviv University, Israel

Edith Boxman

EW JERSEY · LONDON · SINGAPORE · BEIJING · SHANGHAI · HONG KONG · TAIPEI · CHENNAI · TOKYO

Published by

World Scientific Publishing Co. Pte. Ltd.
5 Toh Tuck Link, Singapore 596224
USA office: 27 Warren Street, Suite 401-402, Hackensack, NJ 07601
UK office: 57 Shelton Street, Covent Garden, London WC2H 9HE

Library of Congress Cataloging-in-Publication Data
Names: Boxman, R. L. | Boxman, Edith Selina.
Title: Communicating science : a practical guide for engineers and physical scientists /
   Raymond Boxman (Tel Aviv University, Israel), Edith Boxman.
Description: New Jersey : World Scientific, 2016.
Identifiers: LCCN 2016024823| ISBN 9789813144224 (hardcover : alk. paper) |
   ISBN 9789813144231 (pbk. : alk. paper)
Subjects: LCSH: Communication in science. | Technical writing.
Classification: LCC Q223 .B694 2016 | DDC 808.06/65--dc23
LC record available at https://lccn.loc.gov/2016024823

British Library Cataloguing-in-Publication Data
A catalogue record for this book is available from the British Library.

Copyright © 2017 by World Scientific Publishing Co. Pte. Ltd.

All rights reserved. This book, or parts thereof, may not be reproduced in any form or by any means, electronic or mechanical, including photocopying, recording or any information storage and retrieval system now known or to be invented, without written permission from the publisher.

For photocopying of material in this volume, please pay a copying fee through the Copyright Clearance Center, Inc., 222 Rosewood Drive, Danvers, MA 01923, USA. In this case permission to photocopy is not required from the publisher.

Typeset by Stallion Press
Email: enquiries@stallionpress.com

Printed in Singapore

Preface

Ray Boxman: The unlikely story of this book begins with my early elementary school education in the days when students were issued stubby pencils without erasers and paper that seemed to have hunks of wood sticking out of it. I was known as "Mr. X" because of all of the corrections I needed to make. Nonetheless, I was accepted by each of the universities to which I applied, except the one that required a writing sample. Upon completion of my PhD thesis, the more junior of my supervisors reviewed the thesis and commented that he liked it, but suggested that it should be written in the present tense. I dutifully revised the text and submitted it to my senior supervisor, who also commented that he liked it, but suggested that it should be written in the past tense. In those days, way before the era of personal computers and word processors, revisions involved a great deal of retyping on a manual typewriter. I solved the dilemma by locking both professors in one room, and announcing that I would not release them until they agreed on the proper tense.

After receiving a faculty appointment at Tel Aviv University, I improved my writing, with some helpful mentoring by senior faculty in my department, notably the late Prof. Enrico Gruenbaum (whose native language was not English). I published many papers and thus did not perish in academia.

In 1997, the Dean of the Faculty of Engineering presented me with a problem. Recently, the faculty had changed its regulations to require all Ph.D. students to submit their theses in English. The motivation was to widen the basis for finding referees. However, these referees complained about the level of the English.

The Dean didn't know about my past identity as Mr. X. Also, his motivation in tossing this hot potato to me may not have been totally pure. Although I had certainly published and edited, and English is my native language, I had no prior experience in teaching English or writing. The university had a foreign languages division which previously had taught the faculty's Ph.D. students, but the faculty paid for this service. Because of interdepartmental squabbling about the right to teach various electrical engineering courses, I had a light teaching load, so throwing me the hot potato would save the faculty some money.

I accepted the challenge and examined some of the students' work, and the referees' comments. I quickly concluded that while there was an "English" problem, there was a much bigger underlying problem of poor organization and poor writing that was often connected to poor scientific thinking. The result in any language would be poor.

The course I organized emphasized the organizational aspects of writing theses and journal papers, using Weissberg and Buker's excellent text *Writing-up Research*. Much to my surprise, the students loved the course, as did their thesis supervisors. Frequently, I received feedback from the students that this was the most valuable course in their doctoral studies. Students reported comments from reviewers that their papers were well written and well organized. Some students even received best-paper awards. Furthermore, the students' thesis supervisors felt that their burden was lessened, and some even reported that they learned useful hints from their students. I taught the course for sixteen years, until I reduced my teaching load nearing retirement. Perhaps the Dean's decision to throw this hot potato to an engineer and the former Mr. X had some merit.

Edith Boxman adds: My introduction to scientific writing and editing was proofreading Ray's doctoral thesis. I swore I would never again read any scientific paper by Ray or anyone else. However, in my own work, as an economist and banker, I was frequently asked to translate and edit documents written by my colleagues. These requests, I could not refuse, especially since many of these colleagues were kind enough to edit the documents I wrote in Hebrew. My threat in the early years of our marriage never to review a scientific paper again was forgotten as I wrote business plans for some of Ray's commercial ventures and became increasingly involved in developing materials for supplemental writing courses.

And from us both: In response to several requests from Ray's colleagues, we jointly developed a 12-hour short course on scientific writing, which we subsequently presented at the Aachen Technical University, Ariel University, Istituto Nazionale Fisica Nucleare – Universita Degli Studi di Padova, Kazan State Technological University, the Leibniz Institute for Plasma Science and Technology, Northwestern

University, University of Canterbury, University of Sydney and Xian Jiaotong University.

Our teaching experience provided the impetus and motivation for this text. We wanted a short text that emphasizes the connection between scientific thinking and writing. We wanted a text aimed at engineers and scientists, written from an engineer's perspective. We wanted a text that every Ph.D. student would want to read before writing his or her thesis. And we wanted a resource to which researchers would return as they advanced in their careers. We hope that this text fulfills at least some of these goals.

We are grateful for all of the help and encouragement we received during the course of writing this book. First and foremost, we thank all of our students. The adage that you learn from your students is certainly true for us. Our students' enthusiasm to learn has inspired us. Many of the examples in this text are based on the exercises of students who gave their permission to use their material anonymously. We thank RB's colleagues in the Electrical Discharge and Plasma Lab, Issak Beilis, Nahum Parkansky, and Dan Gazit, as well as students in the lab, whose work inspired many of the examples in the text. We gratefully acknowledge Orna Hamilis, who prepared the illustrations.

Many colleagues, family members, friends, and former students suggested examples or reviewed all or parts of *Communicating Science*. We thank Andre Anders, Benjamin Boxman, Jonathan Boxman, Lillian Boxman, Ian Falkoner, Joyce Friedler, Evgeny Gidalevich, Rami Haj-Ali, Daniel Haney, Azriel Kadim, Simon Kahn, Michael Keidar, Juergen Kolb, Shalom Lampert, Emma Lindley, Zhiyuan Liu, Boris Melamed, Judith Posner, Daniel Prober, Yossi Shacham, Zhiyuan Sun and Sharyn Weisman, for their useful suggestions, and for catching embarrassing mistakes. We, of course, take responsibility for those that remain.

Raymond and Edith Boxman
Herzliya, Israel, April 2016

Contents

Preface   v

1. Introduction   1
2. Research Reports: Journal Papers, Theses, and Internal Reports   6
3. Submitting a Paper and the Review Process   72
4. Conference Presentations: Lectures and Posters   84
5. Research Proposals   106
6. Business Plans   124
7. Patents   137
8. Reports in the Popular Media   154
9. Correspondence and Job-Hunting   163
10. Writing Well: Organization, Grammar and Style   185

Index   265

Chapter 1

Introduction

The ancient Greek mathematician and physicist, Archimedes (287–212 BCE), is said to have run naked through the streets of his native Syracuse shouting Eureka after he discovered, while bathing, the principle of buoyancy which bears his name. This rather unorthodox method of communicating science and technology was supplemented by the more conventional method of writing up his research. Archimedes' research on mechanics and geometry has influenced the scientific community to this day.

Archimedes was a member of a scientific community which was spread throughout the world known to him. He corresponded with colleagues including Conan of Samos and Eratosthenes of Cyrene. The results of one member of the community provided a basis for new results by others. The tradition of communicating among members of the scientific community facilitates synergy among scientists and has led to a rate of scientific progress which would be inconceivable if each had worked in isolation. Thus global communication among scientists has most importantly an altruistic function — furthering science and technology.

Secondarily, the communication of science and technology within the scientific community furthers the reputation and career of the scientist-communicator. It is not only what you do but also how you present it that influences the opinions of evaluators.

Close links between science and government and the need to communicate science to non-experts are by no means new. In his famous run in Syracuse, Archimedes was not headed to a scientific symposium of his colleagues but rather to his patron, King Hiero II. The king had commissioned Archimedes to solve a very practical problem with diplomatic implications — how to determine the gold content of a gift crown of irregular shape without damaging it.

In modern times, scientists and engineers need to communicate with:
(1) the scientific community, to share their research results and to participate in the global effort to advance science and technology;
(2) the public sector, to gain sponsorship and to influence policy;
(3) the business community, to convert research results into commercial products and services which can benefit consumers, and thereby benefitting investors and business backers; and
(4) the general public, to encourage the support of the taxpayer who ultimately bears much of the cost of scientific research and development.

A research report is a communications channel, analogous to a TV broadcast channel, in which the writer is the transmitter and the readers are the receivers. As in broadcasting, there are hopefully many receivers (readers) but only one transmitter (the writer). The transmission is one-directional; the transmitter has no possibility of receiving feedback, requests for clarification or repeat transmissions. To successfully communicate with the receivers, the transmitter uses a protocol, frequency, power level, etc. suitable to the receiver, which must take into account limitations of the media and the receivers. Considerable resources and expense are expended on the transmitter so that relatively simple, inexpensive receivers can be used.

Similarly, the skilled scientific writer uses conventions which the reader expects and endeavors to present the topic in a way that is easy for the reader to understand. The writer must expend much effort so that the readers' task is easy. Good scientific and technical writing is easy for reader to understand but not necessarily easy for the writer to write.

The communication conventions are very much a product of the culture in which they exist. The style and organization of a paper today in the *Journal of Applied Physics* is decidedly different from that used by Archimedes in the classical Greek world.

This guide is intended to assist researchers and graduate students to write theses and journal papers, to prepare lectures and posters, and to communicate through other genres that they are likely to encounter in their professional careers. It provides "recipes" for these genres which are suitable for today's scientific global village. Grammatical and stylistic suggestions are provided specifically for English, but the organizational guidance is equally applicable throughout western scientific culture.

Good scientific communication requires good underlying scientific thinking. Moreover, the process of communicating can sharpen scientific thinking. This book will emphasize this connection.

The "recipes" provided in this book are not hard and fast rules. Many writers produce excellent papers by modifying their style to suit particular content, or they develop a personal style while remaining within the broad conventions used in communicating science. Others deviate from these rules with poor results and yet their work is published. However, the recipes suggested herein will always produce well-organized documents. They will provide the inexperienced writer with a solid starting point.

Typically, researchers and technologists spend 20% of their time writing research reports, proposals and other communications, but few receive formal training in writing. Furthermore, writing often does not come naturally to scientists and engineers, and is often viewed as an unwanted chore and a distraction from actually performing research and development. Nonetheless, writing is a skill that can be learned. With improved skill, the quality of the communications is improved, the effort required is reduced, and the reputation of both the writer and the writer's institution is enhanced.

For speakers of English as a second language, global communication may seem especially challenging. It may be helpful to realize that English in scientific papers is relatively straightforward compared to that in general literature. Scientific writing uses limited vocabulary and grammatical forms, it is conventional and the emphasis is on clarity and not on linguistic gymnastics.

This book is organized first into chapters devoted to particular writing genres, and then is followed by a chapter devoted to composition, style, and English usage which is applicable to all genres.

Chapters 2–5 present genres used within the research community:
- Chapter 2 presents the "research report", which includes theses, journal papers, and internal company reports. This chapter is central to the book, as writing research reports is the most common communications task for researchers, and the other genres contain analogous elements. The concepts developed in this chapter will be used in the subsequent chapters.
- Chapter 3 discusses the journal publication process.
- Chapter 4 discusses conference presentations, both short lectures and posters, as well as suggestions for getting the most out of conference attendance.
- Chapter 5 presents research proposals. Researchers must write successful proposals to advance into leadership positions.

Chapters 6–9 concern communications genres which connect scientists and engineers with the wider community. Chapters 6 and 7 are intended for scientists and engineers interested in commercializing their research results.
- Chapter 6 presents the business plan, a document which organizes the business approach of entrepreneurs, and facilitates evaluation by potential investors.
- Chapter 7 discusses patents and patent applications.

- Chapter 8 discusses reports in popular media.
  Because business plans, patent applications, and press releases are usually prepared by or with the assistance of professionals, the main objective of these chapters is to explain their content and organization so that the interaction between the researcher and professional patent preparer, business analyst, or publicist will be effective.
- Chapter 9 presents business correspondence, curricula vitae and résumés, as well as some suggestions for job hunting.

Finally, Chapter 10 provides general writing guidelines, including writing strategies, English grammatical structures common in scientific communications, and some tips for non-native English speakers.

Chapter 2

Research Reports: Journal Papers, Theses, and Internal Reports

2.1 Definition and Scope

The research report is a complete report of a scientific finding to the scientific community. It specifies how the finding was obtained in sufficient detail to allow duplication elsewhere. It places the work in scientific context by reviewing previous scientific work and by interpreting the current results in view of the most relevant preceding studies. The research report is the most important instrument by which scientists and engineers communicate the results of their research to their professional colleagues.

The research report is distinguished from the abbreviated research report by its completeness. Abbreviated research reports, which include letters, brief communications, and many conference papers, do not necessarily provide all details needed for duplication or for placing the work in its scientific context. Abbreviated research reports will be described briefly towards the end of this chapter, in section 2.13.

Primarily, the purpose of the research report is to share research results with colleagues. Colleagues may be in a restricted circle in the case of an internal report, or throughout the global scientific community in the case of a journal paper. In so doing, the researcher becomes a member of a team, either a restricted team whose objective is to advance the interests of an institution or a loosely defined global team whose objective is to advance knowledge. Secondarily, a good research report enhances the reputation of the researchers who write the report, and the organization in which they work.

The same basic structure is used for various types of research reports, including theses, journal papers, and internal reports. This structure is applicable across a wide spectrum of fields including engineering, physical sciences, life sciences, and social sciences. All experimental and quasi-experimental research reports (e.g. numerical experiments and public opinion surveys) use the same structure. Theoretical papers use a slightly different structure, and there is more variance in presentation among them, but they also share the same basic principles. This chapter explains the structure of the research report and the content of the various sections. It describes style and grammatical conventions relevant to the various sections of the report. Examples of key sentences, as well as the key conventions, are provided.

2.2 Before Beginning to Write

2.2.1 *Research question*

Good research reports revolve around a research question, the question which is answered by the research report. In some fields such as biology or medicine, the research question is stated formally. In engineering and the physical sciences, the research question does not appear explicitly as part of the paper, but is implicitly stated in the statement of purpose (stage 4 of the Introduction, described in section 2.4.4 below) and explicitly answered in the Conclusions (described in section 2.9).

Most likely, a research question was defined when the research was first proposed. If so, review it to determine if the results obtained in fact answer the question. The question may be revised, as necessary, to reflect more accurately what was actually answered by the research. If a research question was not previously defined, formally state the research question before beginning to write the research report, and use it to guide all aspects of the writing.

The research question should be explicit, succinct, and as broad or (more likely) narrow as the actual results obtained. Formulate it as a grammatical question, i.e. it must demand an answer and be terminated by a question mark (?). Examples of research questions are presented in Table 2.1.

Table 2.1. Examples of research questions (RQs). All of the examples in the left column are structurally correct, and can describe a particular research project. The right column analyzes each example, and, where appropriate, suggests alternatives.

Examples of Research Questions	Comments
How does bias voltage affect the adhesion and interface structure of Ti-Al-N coatings applied to stainless steel substrates using vacuum arc deposition?	Precise RQ, which probably characterizes the research accurately.
What is the maximum number of cores in a multi-core fiber in which it is possible to write a single identical fiber Bragg grating notch using the phase mask writing method?	Precise RQ, which probably characterizes the research accurately. Presumably the answer is a number.
Can we reduce the cost of an Al-Sc-Er-Zr-Si alloy by adjusting the concentration of both Sc and Zr, while keeping good mechanical properties?	Precise RQ, which demands a yes/no answer. Presumably the answer is yes, since proving no is difficult.
Do non-thermal effects contribute to microwave killing of bacteria and viruses?	Precise RQ, which demands a yes/no answer. Presumably the answer is yes, since proving no is difficult.
Which metal-ceramic brazing method produces the strongest junction between alumina (AL_2O_3) and stainless steel using as brazing filler copper, silver and Ticusil®?	Precise RQ. Expect the answer to be the name or description of the best method.
Are atmospheric pressure plasmas able to rupture the cell wall of microalgae to extract their valuable compounds and can they be an energy efficient, cost saving alternative compared to standard methods?	Two RQs, each demanding a yes/no answer. Both are presumably yes, as no would be difficult to prove.
What is the formation mechanism of micro-discharges in atmospheric pressure CO_2 plasmas?	Precise RQ. Answer will be a description of the mechanism.

Examples of Research Questions	Comments
Will plasma treatment with kINPen MED and plasmaONE harm the surface of human skin by overheating the skin surface, damaging the structure of the stratum corneum, affecting beta-carotene content, or exposing the skin surface to excessive UV radiation?	Precise RQ if the two treatments are rigorously defined. Demands yes/no answer. If the treatments are not rigorously defined, proving "no" may be difficult.
How can the Multanova 6F speed radar be calibrated electronically for the full speed range of 30–249 km/hr?	Precise RQ. Expected answer would be a calibration procedure.
How can the efficiency of a CMOS rectifier be improved for transcutaneously-powered implants?	Precise RQ. Expect the paper will either compare various means of improving the efficiency, or the exposition of some particular means.
How can the dielectric breakdown properties of SF_6 and metal vapor mixtures be calculated in the 300–3000 K temperature range?	Precise RQ, but is it the real RQ? The answer to this RQ is the exposition of a theoretical method. But possibly a better RQ would be: *What are the dielectric breakdown properties of SF_6 and metal vapor mixtures in the 300–3000 K temperature range?* The answer to the latter RQ would be graphs of the breakdown field as a function of T for various mixtures.
Are supermassive black holes formed by the merging of several intermediate-mass black holes?	Was merging the only hypothesis examined? If not, a better alternative might be: *How are supermassive black holes formed?*

2.2.2 Principal claims

Many authors find it useful to list the principal claims or principal results of their research as an aid to organizing their paper. These claims or results, taken together, should lead the author and the readers to the principal conclusion of the paper, which is the answer to the research question described in section 2.2.1. Generally, each claim or result will be embodied in a figure or table in the Results section of the paper. This exposition of results forms the core of the paper.

2.2.3 Outline

A detailed outline of the research report indicates each section, sub-section, sub-sub-section etc. down to the level of the individual paragraphs (i.e. a separate line entry composed of one or a few words for each paragraph). An example is presented in Table 2.2. Writing a detailed outline is useful in several respects. It guides the writing process by forcing the author to first think about the big issues and not get hung-up on phrasing. And it helps prevent one of the most common problems – misplaced statements (e.g. placing a result of the current research in the Literature Review, or providing a detail of the experimental procedure only after the relevant result is reported).

The outline is recorded as bits and bytes in a word processor, not chiseled in stone. Thus, during the course of the writing it may be revised as the author's ideas become better formulated. Careful consideration of revisions in the context of the outline will help keep the structure logical and avoid add-ons which do not develop the key points.

2.2.4 Considering the reader

The overriding consideration in organizing the structure of a research report is to provide information to the reader in the most easily absorbed form. This implies several operative guidelines:
(1) Use conventional organization as outlined in section 2.3. The reader expects it and can absorb the information most readily if the paper is conventionally organized.
(2) Order the paper so that it most efficiently transmits information to the reader. This is not necessarily the order in which the researcher performed the research. The order taken by the researcher is only relevant if it influenced the results, e.g. the sequence of steps in an experimental procedure.

Table 2.2. Sample outline for a research report

DECOMPOSITION OF DISSOLVED METHYLENE BLUE (MB) IN WATER USING A SUBMERGED ARC BETWEEN TITANIUM ELECTRODES

Introduction
    (Stage 1) Need for new plasma water treatment
    (Stage 2) Background
        Plasma water treatment
        Submerged plasma
        Pulsed submerged arc
    (Stage 3 Gap) – no previous submerged arc treatment + aging
    (Stage 4 Objectives)
        Demonstrate MB removal
        Determine influence of discharge parameters, H_2O_2 addition
        Determine energy consumption

Experimental Details
    Arc treatment apparatus
    Arc treatment procedure
    Monitoring MB concentration
    Examination of solutions and particles

Results: Effect of arc processing on MB concentration
    Arc processing of MB aqueous solution without H_2O_2
        Absorption spectrum
        MB concentration vs time
    Arc processing of MB aqueous solutions with H_2O_2
        Absorption spectrum
        MB concentration vs time
    Influence of particles eroded from Ti electrodes on the MB removal
        TEM images of particles
        Aging experiment
    Influence of filtration and addition of TiO_2 powder on the MB removal
        Filtration experiment – MB concentration vs. aging time
        Addition experiment – MB concentration vs aging time

Discussion
    Species produced by discharge
    Discoloration, reversible
    Discoloration, irreversible
    Aging process
    H_2O_2 addition
    Energy yield

Conclusions
    Arc treatment + aging effective
    Treatment effectiveness increased with pulse energy and H_2O_2 additions
    Ti nano-particles eroded from electrodes by arc – significant positive influence
    Energy consumption least among plasma treatment methods

(3) Design the sequence of presentation to convey information — not to make a good story. The research report is not a murder mystery! Do not withhold facts to keep the reader in suspense for a surprise ending. Order the information so the reader can most readily absorb it.
(4) Spare the reader lengthy descriptions of the authors' thought processes and blind alleys. Concentrate on what was done, and what was observed. The reader is mostly interested in the research, not the researcher.
(5) At every stage of the report, ensure that the reader has been provided with all the information needed to understand that stage.

Who are your readers?

The readers of a journal paper may include practitioners and fellow researchers from your specialty and other fields, and students. Thesis examiners are usually experts in your field, but some may be from a related field. The thesis may be read also by new students entering the field. An internal report will be read by your immediate manager and your colleagues in the group, but it might also be read by upper management.

In each case, some of the readers are apt to be from other fields, or new to your field, and their needs must be taken into account when writing a research report. The Background section of the report should be broad enough so that all readers, including those who are not experts, can understand the body of the report. Descriptions of apparatus and methods, and especially derivation of equations, should be written so that they may be understood and duplicated by all readers. In general, reports should be written so that a beginning M.Sc. student can understand all of the detail.

2.3 Overall Structure of the Research Paper

The conventional structure of an experimental research paper is summarized in Table 2.3. The Abstract provides an overall informative summary. Each section of the main part of the paper answers a basic question.

Table 2.3. Principal sections of a report and the questions which they answer

Section		Content or question answered
Abstract		Short, stand-alone and informative summary.

1 Introduction
    <u>Required</u>
        Background
        Literature Review
        Gap
        Statement of Purpose
    <u>Optional</u>
        Value Statements
        Preview
		What are we talking about?

2 Methodology	BODY	What did we do?
3 Results		What did we get?
4 Discussion		So what?
5 Conclusions		What is the answer to the Research Question? ~3 key points you want reader to remember.

The structure above may be portrayed graphically as two truncated cones connected by a narrow column, as shown in Figure 2.1. The width of the structure indicates the breadth of the coverage. The Introduction begins broadly in the Background section, and narrows to the Statement of Purpose. The focus is narrow in the body of the paper where

Experimental Details and Results are presented. The Discussion begins with a narrow focus, and then broadens as the implications of the work are discussed.

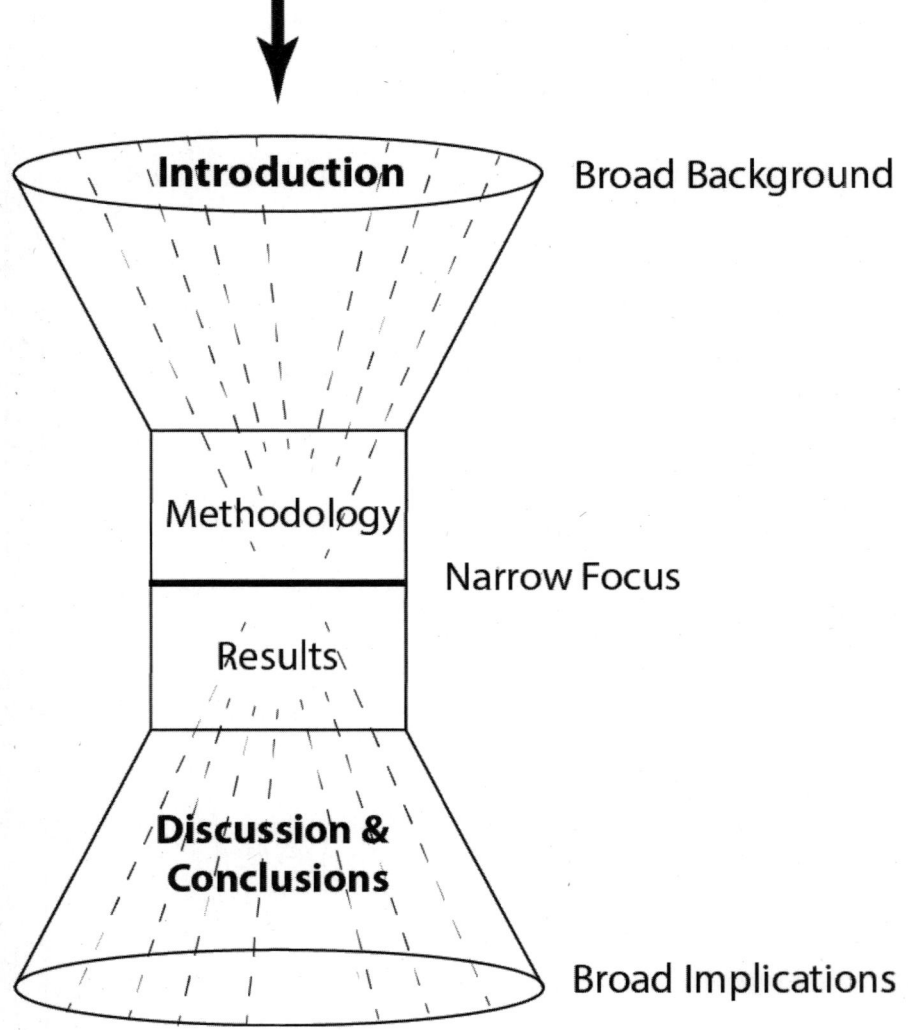

Figure 2.1. Conical representation of the coverage of sections of the research report.

> ## Combined "Results and Discussion"?
> ## A minefield to be avoided
>
> Many authors write papers with a combined "Results and Discussion" section. Our opinion is that all writers, but especially novice writers, should avoid this structure. Results convey information that is wholly objective, and totally repeatable. Every investigator, anywhere in the world, should be able to duplicate these results by following the recipe given in the Methodology section.
>
> The Discussion converts this information to knowledge, by interpreting, explaining, and extrapolating the results. Except in the rare case (in every field except mathematics) where a conclusion can be proved, the Discussion will be somewhat subjective. Given that information is often incomplete, different researchers might legitimately interpret the results differently.
>
> Because of the great difference in their nature, Results and Discussion must be presented differently. Their placement and style must convey to the reader which is which. Results should be conveyed factually, while interpretations, explanations, and extrapolations must be presented modestly and tentatively.
>
> Furthermore, the interpretation and understanding of a phenomenon often depends on the totality of the results. In such cases, all of the facts should be presented to the reader before attempting to interpret them.
>
> Combining Results and Discussion in a single section tends to blur the difference between results, and their interpretation. Many writers using this organization fail to differentiate between the two, and misleadingly present their interpretation as if they were facts. Separating the Discussion from the Results not only declares to the reader which is which, but also helps the writer to use the correct style with each.

2.4 Introduction

The Introduction gives the reader the perspective needed to understand and appreciate the research work. It contains six stages, four mandatory and two optional, each of which will be described in detail in the following sections.

2.4.1 *Background: overall setting*

The Background presents the general context of the research and broadly defines its topic. Its starting point depends on the intended audience. If some of the audience is not likely to be well acquainted with your specific area of research, the Background statement should be more general than if all of the audience are specialists on the topic.

- For a journal article, the Background section should be understandable to the usual readership of the target journal. The introduction may start at a relatively advanced point in a specialized journal such as the *Journal of Ferromagnetics*, but would need to be more general for the *Journal of Applied Physics*, and even more so for *Science*.
- For a thesis, it must be considered that one or more of the examiners may not be specialists, and that subsequent students continuing the research may use the thesis as a primer.
- For an internal report, where everyone on the distribution list is intimately involved in the area of research, this stage can be quite short. However, if senior managers are among the intended audience, then the Background should be broadened accordingly.

The introductory sentence, i.e. the first sentence of the Introduction, should provide a clearly understandable and informative overview to the broadest relevant audience. Avoid non-informative generalities e.g. *It is an important problem in physics* or *It has many applications*. Instead, briefly describe the problems and/or give examples of the applications. Table 2.4 presents examples of good introductory sentences.

Table 2.4. Examples of introductory sentences.

Examples of Introductory Sentences	Comments
The waste effluent from textile, leather, food processing, dye manufacturing, cosmetics, and paper industries includes dyes that pollute the environment if untreated.	States the general problem. Suitable in a broadly aimed journal.
Purifying water from chemical and biological contaminants is a challenging scientific and technological problem	States the overall technological problem. Suitable in a broadly aimed journal.
Vegetation, such as trees, shrubs and grass, plays an important environmental role in depollution, storm water control, and energy saving.	Defines the general field. Suitable for a broadly aimed journal.
Applying optical fiber technology to astronomical instrumentation allows the development of small, cheap and versatile devices.	Presents advantages of technology which is the focus of the paper. Suitable in a specialized journal.
Plentiful observational evidence supports the existence of supermassive black holes (weighing a million times more than our sun) at the centers of most large galaxies; however, their formation mechanisms are not well understood.	Presents the open question. Suitable in a specialized journal.
Brazing is a welding technique whereby pieces are joined by a filler metal which has a fusion temperature above 450 °C, but below the fusion temperature of the brazed piece.	Defines the general topic.
The number of medical procedures involving the use of cyclotron-produced radionuclides is growing year by year.	Presents the application and implies its growing importance.
The metallic plasma generated by the vacuum arc has various technological applications including depositing thin films and coatings, filling high aspect ratio trenches on Si chips, and implanting ions.	Presents applications of the technology discussed in the paper. Suitable in a specialized journal.
In order to reduce oil consumption, the transport industry aims to reduce vehicle weight by introducing new low-density materials such as aluminum alloys.	Presents the incentive for studying this field.

The Background stage is normally one paragraph, containing 3–5 sentences. Usually, the sentences are very general and non-controversial, starting from a general description of the topic and ending with a reference to the key issue of the present report. Within each sentence, old information should be provided first, and then new information. An example of an opening paragraph is presented and analyzed in Table 2.5.

Table 2.5. Example of an opening paragraph in the Background section.

Text	Comment
New water treatment methods are needed to supply safe drinking water to an expanding population, and to treat waste water before it is discharged into our increasingly fragile environment.	Broad opening statement, understandable by all readers. Explains motivation. Uses familiar, non-technical words.
One family of new water treatments uses plasma, the energetic fourth state of matter in which constituent atoms and molecules are ionized.	*Treatment* is old information because it was introduced in the previous sentence. The new information, *plasma*, is defined.
The plasma may be used to kill bacteria and decompose organic contaminants, using a combination of UV radiation and oxidizing radicals created in the plasma.	Develops the concept of *plasma* (now old information), and presents the underlying mechanism of plasma water treatment (*UV, radicals*).
The plasma may be generated above the water, or using a submerged electrical discharge, within bubbles in the water.	Further develops *plasma*, introduces new information – *plasma bubbles*.
Plasma bubble treatment is particularly effective, because all of the UV radiation and oxidizing radicals are absorbed by the surrounding water.	Conclusion, combining informational elements introduced earlier (*treatment, plasma, bubble, UV, radicals*).

2.4.2 Literature Review

The Literature Review places the present research in the context of relevant published work. It shows referees that the author is familiar with the field. Together with the following two stages of the Introduction (the Gap and the Statement of Purpose), it demonstrates the uniqueness of the present work. This is usually the longest section of the Introduction. It is generally composed of sentences which either cite reviews of previous works or specific works.

Citations are typically grouped or ordered according to one of the following methods, chosen according to the specific needs of the paper:

- relevance, starting with the least relevant, and ending with the most relevant;
- approach, beginning with the approach furthest from that used in the present work, and ending with the closest; or
- chronology, beginning with the earliest, and ending with the latest.

The extent of the Literature Review varies with the type of report. In internal reports, the citations are often limited to the most relevant. In journal papers, citations should be chosen to firmly establish the context of the present work. The extent of the Literature Review in a thesis will be determined by the policy of the university or thesis supervisor. In some cases, it may be similar to a journal paper. In other cases, the Literature Review is considered an essential element of the student's training in its own right, and a much more extensive review is demanded. The student might be instructed, for example, to comprehensively review the history of the field or to critique the most relevant papers more deeply.

Various types of citation sentences are used, according to where they appear in the literature review. The Literature Review typically begins with citations defining the general area of the study. Multiple references are usually cited. Such citations are typically information prominent — i.e. some aspect of the information itself is the grammatical subject of the sentence. Some examples are given in Table 2.6.

Table 2.6. Examples of information prominent citations. The subject of each sentence is underlined.

Because of the complexity of the non-equilibrium behavior, <u>the swarm parameters</u> have been analyzed in non-uniform fields in He [1], [2], and N_2 [3], [4] by Monte Carlo simulation and in air and Ar [5] by solving the diffusion flux equations.
<u>The supermassive black holes</u> with masses between 10^6 and 10^9 solar masses, are found in the nuclei of most nearby massive galaxies [1] [2].
<u>Current density on the anode surface</u> has been investigated by a number of researchers [1-4].
<u>Sodium</u> is one of the most effective dopants in controlling hole carriers in IV-VI thermoelectric alloys and has been extensively utilized to optimize their thermoelectric properties [1].
As reported by Sublet, et. al [1], <u>the temperature at which niobium is deposited</u> influences niobium thin film morphology and hence its RF-properties. <u>The ideal deposition temperature</u> is between 400°C and 600°C.

> *Chemical vapor deposition (CVD) can synthesize scalable MoS$_2$ monolayers, using MoO$_3$ and MoCl$_5$ as precursors [1].*
>
> Note: in the last three examples, although a specific work is cited, *sodium, temperature* and *CVD*, respectively, are the subjects. This makes these citations "information prominent" rather than "author prominent".

An alternative form typically used early in the Literature Review is weak author prominent. In this form, a phrase such as *several authors* or *many authors* is the grammatical subject of the sentence. Such sentences often describe typical work performed in the past. Some examples are given in Table 2.7.

Table 2.7. Examples of weak author prominent citations. The subject of each sentence is underlined.

> *Several authors [1-4] have studied linear and nonlinear wave processes in various kinds of plasma, but not much work has been done for waves in a rotating plasma.*
> *A few researchers [1-5] have suggested using non-thermal plasmas to convert CO$_2$ to more valuable hydrocarbons.*

Single references are often cited as the Literature Review progresses. The citation focus is typically author prominent when the studies described are most closely related to the present work. The cited author is the grammatical subject of the sentence. Some examples are presented in Table 2.8.

Table 2.8. Examples of author prominent citations. The subject of each sentence is underlined.

> *Boeuf et al. [6] took into account the additional ionization caused by a beam-like group of fast electrons and developed an extended memory factor model in helium.*
> *The GNOSIS project [1] demonstrated the effectiveness of fiber Bragg gratings in removing atmospheric emission lines from near-infrared spectra.*
> *C.C. Souza et al. (1) studied the average effect of grain boundaries in samarium doped ceria and reported that nano-structuring lowered the ionic conductivity monotonically when grain size was decreased from 73 to 33 nm.*
> *Donovan et al. [4] discovered that people experienced more deaths from heart and respiratory diseases when they lived in areas without trees.*
> *Lebouvier et al. [6] discussed how CO$_2$ dissociation could provide a new route for Syngas production.*

Towards the end of the Literature Review, the state of the art might be summarized. Some examples are given in Table 2.9.

Table 2.9. Examples of state of the art summary citations. The subject of each sentence is underlined.

<u>Sputtering</u> was found to be an efficient technique to produce amorphous metallic alloy coatings [22-24], because it provides substantially higher quenching rates than conventional quenching techniques.
The <u>synthesis of fluid vesicles</u> from various surfactant molecules has been extensively studied experimentally [3, 4, 6-10] and theoretically [106].
<u>Phase-locked oscillators</u> have been studied for nearly 80 years and have been shown to provide a stable and controllable frequency [4-13].

Each paragraph in the literature review should have at least one citation. If only one citation is found in a paragraph, the reader will assume that everything presented in that paragraph is attributed to that single citation. It is the authors' job to ensure that this assumption is justified.

Citation symbols (e.g. [3]) should not be used as words in a sentence. If the number must be "said" for the sentence to be grammatically complete, the sentence must be rephrased. It is grammatically correct to write *Reference 3*, but it is much better to refer to the author by name. Using *Reference 3* forces the reader to divert attention from the text and to find the list of references, which are most likely on another page, in order to identify the author of Reference 3. Furthermore, numbers are not usually remembered by the reader. The author's name is easier to remember and gives immediate meaning to the reader who recognizes the name.

The verb tense chosen in informative sentences conventionally indicates the authors' attitude towards the work. The present tense indicates a generally accepted scientific fact, while the past tense indicates a specific, often experimental, result which might not have general applicability. Modal auxiliaries (e.g. may, might or could + the past participle) are used to indicate tentative or doubtful results. See the examples in Table 2.10.

Table 2.10. Author's attitude indicated by tense.
*Note that the last three examples convey the same information (pressure stabilizes), but the choice of tense conveys the author's attitude about the information and thus guides readers' interpretation of the information.

Context	Tense	Examples
Scientific fact	Present	It *is* generally accepted that bacteria primarily multiply by asexual division.
State of the art, or commonly accepted information	Present	Interferometry *is used* to measure small displacements
Result which the author is indicating to be generally true	Present	Jones found that increasing the pressure *stabilizes* the process.*
Specific result which may not have broader implications	Past	Jones found that increasing the pressure *stabilized* the process.*
Tentative or doubtful result	modal auxiliary	Jones indicated that increasing the pressure *may have stabilized* the process.*

2.4.2.1 Citation style

The style of the citation in the list of references is dictated by the individual journals or the university graduate committee. Some require lists numbered by order of citation in the body of the paper; others specify lists alphabetized by author. Some require only the author, journal, volume, starting page number, and year, while others also require the title of the article and inclusive page numbers. Journals publish style instructions on their websites. Conferences typically issue author style instructions to ensure uniformity among papers in the proceedings. It is advisable to check the requirements of the target publications before searching the literature, so that all required information is gathered. Authors can save considerable time using one of the various citation manager programs available online or as an intrinsic or add-on feature in various word processors.

2.4.3 The Gap or the need for the present work

The Gap statement summarizes the state of knowledge by indicating:
- what was not done,
- errors in previous work, or
- disagreement or controversy between various sources.

It is **the most important statement** for getting a paper accepted. A very common cause for paper rejection is that the referee feels that nothing new was presented. The Gap, by indicating what wasn't done previously, shows that the present work is new. A good Gap statement forces reviewers to work hard to reject a paper for lack of novelty.

The Gap statement is usually one sentence. This statement should be **explicit, precise,** and **focused**. It is the transition between the Literature Review and the Statement of Purpose. It often starts with a connective word which signals this transition. Connective words commonly used include:
- *although,*
- *however,*
- *despite,*
- *though,*
- *but*

The Gap sentence is always negative. Negative words commonly used include:
- *no*
- *not*
- *never*
- *no one*

The Gap must be explicit and not be "wishy-washy" (i.e. vague or indecisive). It is inadequate to state *"To the best of our knowledge, no one has...."* It is the authors' job to know the literature and to be in a position to say definitively whether the topic has been investigated or not. Similarly, it is not sufficient to say *"Few researchers have investigated...."* This begs the question – what about the few who have investigated your area of research? These few presented the most relevant previous works. Those works should be the focus of the literature review and the Gap should be stated relative to them. Some examples of wishy-washy Gap sentences and how to fix them are given in Table 2.11.

Table 2.11. Examples of wishy-washy gap sentences are listed below under "Poor phrasing."

Poor phrasing	Better phrasing
However, few quantitative data regarding the influence of the magnetic field on the current density on the anode surface have been reported.	*However, quantitative data regarding the influence of magnetic fields stronger than 0.5 T on the current density on the anode surface have not been reported.*
To the best of our knowledge, the influence of strong 2.45 GHz radiation on E. coli *viability has not been reported.*	*However, the influence of strong 2.45 GHz radiation on* E. coli *viability has not been reported.*
Only a few ultraviolet radiation sensors have been fabricated using zinc oxide or zinc oxide related materials.	*Ultraviolet radiation sensors have not been fabricated using zinc oxide or zinc oxide related materials with responsivity or reliability comparable to commercially available sensors.*

Previous papers by the authors and the authors' groups should be treated in the same manner as other papers. When a paper reports a continuation of previous research, the gap must be defined also in relation to the group's previous work. The Gap sentence defines the need for the present paper rather than for the whole line of the group's research.

How can you be sure that you have a gap?
By searching the literature thoroughly!

Many writers doubt their own knowledge of the subject, and express their uncertainties by writing a "wishy-washy" gap sentence. The lack of a definitive gap sentence detracts from the paper and casts doubt on the knowledge of the authors.

It is the authors' job to know their field thoroughly. They must search for and read the most relevant past literature. They must follow new developments. The literature search process begins with the introduction of the researcher to the topic. Often the topic is assigned or suggested to a junior researcher by a supervisor. Most often the topic is the continuation of ongoing work in the supervisor's group, in which case there are previous research reports from the group. Less commonly, a new topic is motivated by a new report by a researcher not associated with the group. These reports serve as the starting point of a thorough literature search. Read them first to become acquainted with the terminology used in the topic area and to assemble a list of search terms. Then read the references cited in these initial reports, with the aim of finding any alternative version of these search terms. Search for papers which cite these initial reports (using the *Science Citation Index* or *Web of Science*). Initially, at least, also look for alternative versions of the search terms. Reference librarians in the researcher's library can assist in the search process and usually are glad to instruct new researchers in the art of literature searching.

Equipped with a list of search terms and their variants and alternatives, search the literature with a specialized research-oriented web search tool, such as *Web of Science*, to located published papers and books containing these search terms. General web search tools, such as *Google*, may be used to supplement the search, but should not be relied upon. Once a list of papers has been assembled, these papers may be read cursorily, e.g. titles and abstracts only, to see which are relevant, and which are not. Read the relevant papers in their entirety. Finally, check the references in the relevant papers, and locate additional papers citing these papers and if relevant, read them. Repeat the process of checking references, and papers citing relevant papers, until no new relevant papers are found.

Ideally, searching and reading the literature is the first step in a research project, and the knowledge derived from the search is used to shape the research. However, reading the literature should continue throughout the research project. Many researchers find it best to schedule a weekly time to scan the most recent editions of the principal journals carrying papers in their field, and read both the relevant papers and new papers on related topics.

Although you can never be 100% certain that every possible paper has been examined, by following the above procedure, you learn the field, and can confidently write a definitive, focused gap sentence.

If the gap is an error in the previous work of others, it is very important to phrase the gap sentence politely and diplomatically. Previous researchers deserve the courtesy that you would want from other researchers following you. Some examples are given in Table 2.12.

Table 2.12. Examples of a gap due to errors in previous work.

Aggressive phrasing	More diplomatic phrasing
While Shulmann claims to have measured the electron temperature spectroscopically, his determination incorrectly assumed local thermodynamic equilibrium [22]. This unjustified assumption was based on the erroneous identification of the plasma lifetime with the discharge duration of several milliseconds, whereas the plasma lifetime is correctly identified with the plasma transit time of ~1 µs, from the cathode spot where it is created, to the anode upon which it condenses.	*The only previous electron temperature measurement [22] used a spectroscopic technique that requires local thermodynamic equilibrium, which was justified by equating the plasma lifetime with the discharge duration of a few milliseconds. It is now understood that the plasma lifetime is determined by the transit time of ~1 µs, from the cathode spot where it is created, to the anode upon which it condenses, which is insufficient to achieve equilibrium, and thus a different measurement technique must be used.*
Newall [8] erroneously determined the electron drift velocity, and from it under-calculated the electron density, by equating the energy absorbed by the anode (measured by its temperature rise) with the directed kinetic energy of the electrons. Based on this mistake, Newall describes an entirely wrong scenario of a collision-less plasma, where in fact the plasma is collision dominated.	*A previous model of the plasma [8] was based on overestimated electron drift velocity, resulting in an underestimated electron density, producing a collision-less plasma. In light of recent electron density measurements [4], a new collision-dominated model is needed, as well as reinterpretation of the anode temperature rise experiment.*

2.4.4 *Statement of Purpose*

The Statement of Purpose (SOP) presents the objective of the research or the paper, which is to fill the previously stated gap and to answer the research question. It immediately follows the Gap.

The statement should be concise, precise, and focused. The objective of the research is **not** to **do research** (or its equivalent, e.g. *to study, investigate, examine, observe, analyze*). Such statements suggest that the authors cannot place their work in context. Instead, more decisive terms should be used, e.g. *measure, determine, construct, calculate*.

The objective should not be stated as reaching some goal by some specific means, e.g. *The objective of this research was to measure the electron density by laser interferometry.* If the objective was to measure the electron density, a rational researcher would choose the best available means. The objective should be clearly stated here; the means should be described in the Methodology section (see section 2.5). Writing a Statement of Purpose which confuses the objective with the means to reach that objective often indicates that the researchers are confused about what their objective really is. For example, the real objective might be to determine the properties and subsequent advantages or disadvantages of the proposed means with respect to other means. Or possibly the advantages of the means are well known, but they have not been used in some application, and the real objective is to determine if they can be successfully used in that application. Some examples of problematical Statements of Purpose, and how to fix them, are offered in Table 2.13.

Table 2.13. Fixing two problematical types of Statements of Purpose.

Poor Phrasing	Better Phrasing	Explanation
The objective of this work was to increase the current emitted by the cathode by applying carbon nanotubes on the cathode surface.	The objective of this work was to increase the current emitted by the cathode. or The objective of this work was to determine if the current emitted by the cathode can be increased by applying carbon nanotubes on its surface.	Don't state the purpose as reaching a goal by some specific means.
The objective of this investigation [or work, project, study] was to investigate [or study, observe, analyze] the herding behavior of African elephants.	The objective of this investigation was to determine which factors strengthened or weakened the herding tendencies of African elephants.	Don't state that the purpose of the research was to do research.

The research question should be implicitly clear from the Statement of Purpose. Generally, the research question can be derived from a properly stated Statement of Purpose, or *vice versa*, by a simple verbal transformation involving slight rearrangement of the words in the sentence.

The Statement of Purpose may either focus on the research, or the report. A "research" focused Statement of Purpose uses the past tense, e.g. *The objective of the (research, project, investigation, etc.) was* .... In contrast, a "report" focused Statement of Purpose uses the present tense, e.g. *The objective of this (paper, report, article, etc.) is*.... Examples of each type of Statement of Purpose are given in Table 2.14.

Table 2.14. Examples of Statements of Purpose and corresponding research questions.

Focus	Statement of Purpose	Implied Research Question
Research	The objective of this research was to determine quantitatively the speed of the space charge expansion.	At what speed does the space charge expand?
Research	The objective of this work was to determine quantitatively the influence of the magnetic field on the current density on the anode surface.	What is the distribution of current density on the anode surface and what is the influence of the magnetic field on the current distribution?
Research	The objective of the current work was to determine if amorphous binary alloy coatings could be obtained by co-sputtering of the refractory metals Nb, Ta, and Zr and to determine the impact of amorphization on the barrier properties of these coatings.	(1) Can amorphous binary alloy coatings be obtained by co-sputtering of the refractory metals Nb, Ta and Zr? (2) What is the impact of amorphization on the barrier properties of these coatings?
Report	The objective of this paper is to model the relationship between the Al content in Al-Cu alloys and their hardness.	What is the relationship between the Al content in Al-Cu alloys and their hardness?
Report	The objective of this paper is to derive a relationship between macromolecular size and permeability through ceramic membranes.	What is the relationship between macromolecular size and permeability through ceramic membranes?

2.4.5 *Value Statement – justification of the present work (optional)*

A Value Statement indicates the importance or significance of the work. It should be short (1–2 sentences), specific and phrased modestly. A Value

Statement is optional. It may be unnecessary if the value of the work is obvious. However, it can be useful to consider the benefit that your research provides, even if you choose not to explicitly state it in your paper. Being clear about the value of your research can be very helpful in defining your research objective. In internal corporate reports, it may help convince decision makers of its relevance to business objectives.

Modal auxiliary verbs (e.g. can, could, may, might + past participle) are frequently used to express modesty in Value Statements. The Value Statement should not be used to explicitly claim novelty – novelty is implied already by the combination of the Gap and the Statement of Purpose. Some examples of Value Statements are given in Table 2.15.

Table 2.15. Sample Value Statements.

Example	Comment
The use of an appropriate projective cooling can decrease drastically the amount of long-lived impurities during ^{18}F production and thus increase the labeling yield and the specific activity of the final radio-medicine.	States advantages.
Filters based on multi-core fibers are faster to produce, less susceptible to manufacturing variations and more compact than existing filters based on arrays of individual single-mode fibers.	States advantages.
Determining whether the intermediate mass black holes are the building blocks of the supermassive black holes will improve our understanding of the link between the evolution of the black holes and their host galaxies.	Links result to broader context.
If atmospheric pressure plasma rupturing of microalgae cell walls proves to be effective and efficient without altering the extraction products, these products, used as biofuels, food and pharmaceutical agents, could be produced at lower cost.	Explains practical benefit
The results will help industries using brazing to choose the most appropriate methods and improve their product quality.	Explains benefit to industry

2.4.6 Preview (optional)

A Preview may reveal the principal result of the report and/or indicate its organization. It is particularly useful in long reports. Revealing the principal result at this early stage focuses the attention of the reader. If the reader knows the principal result, he will seek the steps needed to reach this result as he reads the paper.

Indicating the organization of the report is not necessary if it is completely standard. In more complex papers (e.g. containing both experimental and theoretical parts) and long works (e.g. theses), informing the reader about the structure is helpful. Examples of preview statements are given in Table 2.16.

Table 2.16   Examples of Previews.

Example	Type
This paper first develops a model for non-linear gas expansion in a chemical explosion, and then tests the model experimentally.	Organization
It will be seen that non-linear effects significantly increase the expansion velocity.	Principal result

2.4.7   Compatibility of the Research Question, Gap Statement, Statement of Purpose and Conclusions

The research question, Gap, Statement of Purpose, and the answer to the research question (in the Conclusions), as well as the Value Statement (when included), must be mutually compatible. One of the most common errors is lack of consistency amongst these elements. Inconsistencies confuse the reader and may indicate that the researchers do not understand the logical relationship between these elements.

- The **Research Question (RQ)** is the objective of the research, expressed as a one sentence question. The RQ must demand an answer, be a grammatical question, and end with "?". Although it is fundamental in defining the focus of the work, it does not directly appear in the paper. The RQ might be stated early in the project as a general statement of intent. But as the project progresses and its objectives are further refined, it is important to revise the RQ accordingly.

- The **Gap** explicitly states what was **not** done previously or what was wrong in the previous work, using a negative word (not, never, etc.). Presumably the gap is that the RQ has not yet been answered.
- The **Statement of Purpose** (SOP) states the objective of the work. The objective should be to answer the research question and to fill the gap.
- The optional **Value Statement** is an outgrowth of all of these elements. It describes the potential benefit of the research. It answers the question: "So what?"
- In the **Conclusions**, the research question should be answered explicitly and informatively, thus filling the gap.

Some examples of compatible sets of these elements are given in Table 2.17.

Table 2.17. Examples of congruent research questions (RQ's), Gap statements, Statements of Purpose (SOP), and answers to the RQ's.

RQ	How do the dielectric properties of pure epoxy and epoxy/crepe composites differ?
Gap	However, no publications have reported on the differences of the dielectric properties of pure epoxy and epoxy/crepe composites.
SOP	The objective of this work was to identify the differences in dielectric properties of pure epoxy and epoxy/crepe paper composites.
Value (optional)	The results could facilitate improved long-term performance monitoring of high voltage bushings.
Answer to RQ (in Conclusions)	Epoxy/crepe paper composites have larger values of real permittivity and imaginary permittivity than pure epoxy, and this is due to a higher moisture content absorbed by the crepe paper.

RQ	How could an option contract influence wind power producer profitability and wind power trading?
GAP	The influence of option contracts on the trading of wind power has not been investigated to date.
SOP	The objective of this paper is to determine the influence of an option contract on wind power producer profitability and wind power trading.
Value (optional)	Adoption of the option contract developed in this work could improve the profitability of wind power producers, promote integration of wind power, and attract more elastic demand for lower-cost wind power.
Answer to RQ (in Conclusions)	Option contracts which pay the producer according to the total energy produced, regardless of availability, maximize wind producer profitability, but they encumber the trading process and lower the demand for wind power.

RQ	What is the maximum number of cores in a multi-core fiber in which it is possible to write a single identical fiber Bragg grating notch using the phase mask writing method?
GAP	The writing of identical Bragg gratings in every core of a multi-core fiber has never been achieved.
SOP	The objectives of this work were to determine the maximum number of cores in a multi-core fiber in which it is possible to write a single identical fiber Bragg grating notch using the phase mask writing method, and to write identical fiber Bragg gratings into this number of single-mode cores in a multi-core fiber.
Value (optional)	Filters based on multi-core fibers are faster to produce, less susceptible to manufacturing variations and more compact than existing filters based on arrays of individual single-mode fibers.
Answer to RQ (in Conclusions)	It is feasible to write matched gratings in a maximum of seven cores to a depth of 30 dB at one time with a 150 mW laser.

RQ	What are the dielectric breakdown properties of SF_6 and metal vapor mixtures in the temperature range 300–3000K?
GAP	Although the dielectric strength of hot SF_6 gases has previously been investigated, the effects of copper vapor were not taken into account.
SOP	The objective of this work was to calculate the dielectric breakdown properties of SF_6 and metal vapor mixtures in the temperature range 300–3000 K.
Value (optional)	The results of this work can improve the design of SF_6 circuit breakers.
Answer to RQ (in Conclusions)	The dielectric breakdown field of SF_6 and Cu vapor mixtures decreased with temperature and Cu vapor fraction.

2.5 Methodology

The Methodology section may be alternatively titled "Experimental Apparatus and Procedure", "Experimental Set-up and Method", "Methods and Materials" (common in chemistry and life sciences), and in theoretical papers (to be discussed in Section 2.7) "Model Assumptions", "Development of the Model", and "Derivation of the Model Equations". All titles must contain a noun or a noun phrase, and thus this section should

not be titled "Experimental". "Experimental" is an adjective and must modify a noun such as "Details", "Set-up" or "Methods".

The Methodology section answers the basic questions: "What did I do?" and "How did I do it?". Part of "how" is equipment, apparatus, materials, samples, procedures, etc

2.5.1 *How much detail is required?*

A fundamental requirement for scientific publication is to provide enough detail so that any other skilled researcher can duplicate the results. All details of apparatus and procedure necessary for obtaining each of the results must be given. Some judgment may be necessary. The choice will generally depend on the author's understanding of what details are essential for duplicating the results. If in doubt, it is best to err on the side of more detail, as reviewers will often reject papers if essential details are missing.

Furthermore, it is recommended to give additional details that will make duplication easier, for example, unusual tricks and procedures that save time. Such detail might include the manufacturer and model numbers of equipment that facilitated performing the experiments or of equipment or material that is especially difficult to find.

However, superfluous detail should be omitted from the text and diagrams. For example, in the text *The arc electrodes were mounted in a vacuum chamber fabricated from 316L stainless steel*, the specification of 316L could be omitted if the grade of stainless steel did not influence the result, and the entire phrase *fabricated from 316L stainless steel* may be omitted if the choice of chamber material did not influence the results. In diagrams, components not critical to the results such as nuts and bolts, O-rings, etc. can and should be omitted.

2.5.2 *Apparatus*

The section on apparatus applies to: equipment, materials, samples, specimens, subjects (laboratory or natural fauna/flora, human subjects), computer programs, algorithms, terrain, questionnaires and surveys. The level of detail required is less for standard apparatus than for special designs.

2.5.2.1 Standard apparatus

Standard or well-known apparatus should be identified, using the past tense. See Table 2.18 for guidelines and examples.

Table 2.18. Identification and description of standard equipment.

Principle	Example
Generic identification is sufficient if any member of that class would yield the same result.	...the data were recorded in a *personal computer*.
Identify characteristics of standard equipment which are believed to influence the result.	The samples were photographed with a *5 Megapixel* digital camera.
If the choice of a particular model might influence the results or could be advantageous for conducting similar studies in the future, the manufacturer and model number should be specified.	The samples were examined with an *Angstrom AA8000-scanning electron microscope*.
If a particular characteristic is important and the choice of a particular instrument is beneficial, then both the characteristic and the model should be specified.	The waveform was recorded with a *Rigol DS1103E digital oscilloscope having a band-pass of 50 MHz*. or The waveform was recorded with a *50 MHz digital oscilloscope (Rigol DS1103E)*.

2.5.2.2 Specially designed apparatus

In contrast, specially designed or unusual equipment must be described and not merely identified. The description should include the following three stages:
(1) overview of the purpose of the apparatus and the overall principle by which it works (typically 1–2 sentences),
(2) description of the principal parts according to some logical order, i.e. spatial arrangement (e.g. top to bottom, left to right, center to outside) or functional arrangement (e.g. in the order in which the

parts operate or process material or information, from the beginning to the end), and
(3) functional description, detailing how the parts work together, preferably in the order in which the parts operate.

Usually the textual description is accompanied by a diagram, especially if the apparatus is complicated. Diagram presentation standards include:
- Using a schematic diagram; neither workshop diagrams nor photographs are suitable because they contain too much detail that distracts the reader from the essence of how the apparatus works.
- Maintaining consistency in nomenclature between the textual description and the labels in the figure, and throughout the report. If a component is called a *screw* in one place, it should be called thus throughout the paper, and never be called anything else, e.g. a *bolt*.
- Including and labelling in the diagram all parts described in the text. All parts appearing in the diagram must either be obvious or be described in the text.
- Ensuring consistency between the orientation and arrangement in the diagram and the textual description. If possible, maintain the orientation and/or arrangement from figure to figure, if the same or similar parts appear in multiple figures.
- Preferably, using **Heads-up Display**[1] giving all the essential information needed to understand the figure, directly on the figure. Label features directly on the diagram with words or obvious abbreviations. An example of a schematic diagram of equipment incorporating the Heads-up Display principle is given in Figure 2.2, and the accompanying textual description is in Table 2.19.

---

[1] Heads-up Display is a device used in combat aircraft that projects important information (e.g. direction, altitude, speed, fuel, ammunition) directly to the pilot's eye, so that he can see this information while looking in any direction. Thereby the pilot can fix his attention on the terrain and enemy aircraft and still receive this information. In the context of a research report, implementation of the heads-up display principle gives the reader all of the needed information to understand a figure without the need to search for information in the caption, or worse, in the body text.

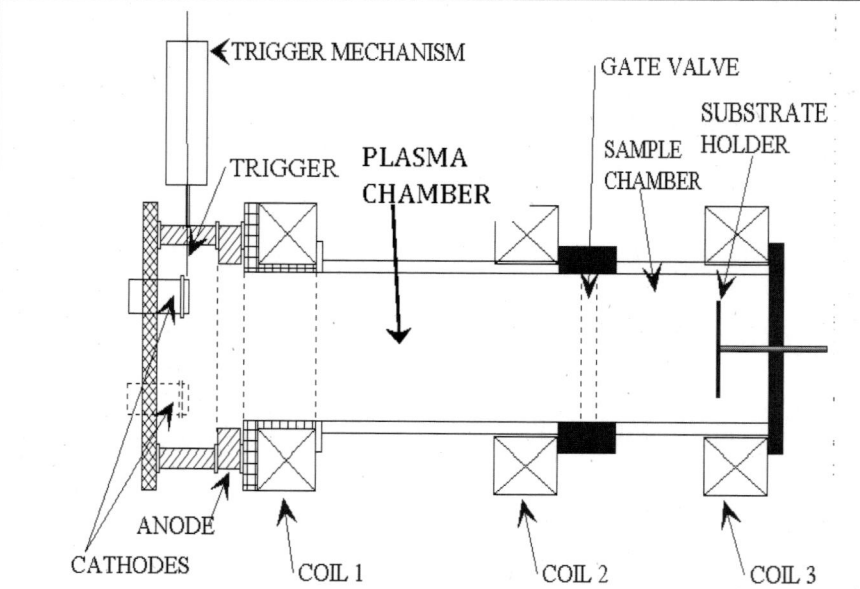

Figure 2.2. Vacuum arc deposition system. [example of an apparatus diagram].

Table 2.19. Sample textual description of the apparatus in Figure 2.2.

Coatings were deposited using a multi-cathode vacuum arc deposition system, shown schematically in Figure 2.2.	(1) overview
Plasma was generated at cathode spots on the surface of three 50 mm diam cathodes, each of a different material, arranged in a circular pattern on the left end flange of the apparatus (two are shown). A trigger mechanism and trigger were mounted adjacent to each cathode (only one is shown) so that the trigger could contact the side of its corresponding cathode. A common 160 mm i.d. annular anode was mounted opposite the cathodes, on the left end of a 160 mm i.d. cylindrical plasma chamber. An axial magnetic field was generated by coils 1–3 both in the plasma chamber, and a sample chamber having the same diameter. The two chambers were connected to either side of a gate valve, which allowed samples to be exchanged while maintaining the plasma chamber in vacuum. Sample substrates were mounted on the substrate holder, which was supported by a flange at the right end of the sample chamber.	(2) principal parts in Figure 2.2 from left to right
Arcs were initiated when the trigger struck and parted from a cathode. Any combination of the cathodes could be operated simultaneously to deposit alloy coatings, or they could be operated sequentially to deposit multi-layer coatings. Metallic plasma composed of evaporated and ionized cathode material, which was generated by cathode spots on the cathode surfaces, passed through the annular opening of the anode, and was directed by the magnetic field through the plasma chamber and sample chamber to the sample substrate, mounted on the substrate holder, where it condensed to form a coating.	(3) functional description, in the order of plasma flow

In contrast, avoid numerical labelling of apparatus components, illustrated in Figure 2.3. Numerical labelling forces the reader to search for the explanation of each part in the caption, or worse, in the body text. Looking back and forth between the figure and wherever the explanation is placed wastes time, annoys the reader and hinders absorption of the information.

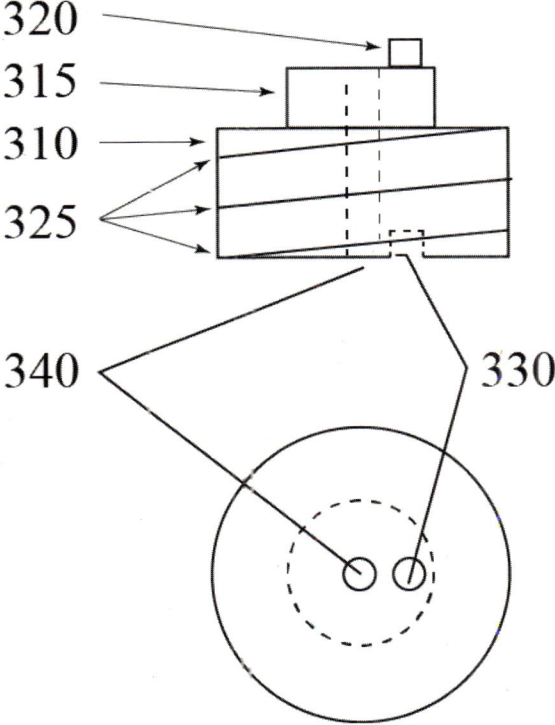

Figure 2.3 Eye-tiring numerical labeling of parts. This type of labeling was common in journals published in the former U.S.S.R. and is used in patent applications, but should be avoided in research reports because it forces the reader to search for explanations of the numbered parts in the caption, or worse, in the body text.

2.5.3 *Procedure*

The Procedure section describes the sequence of steps taken to obtain the results. The format described below is applicable to a variety of procedures including experimental protocols, numerical methods and derivation of equations. Each procedure employed must be described in sufficient detail to allow duplication elsewhere. The various procedures should be described in the order in which the various steps were executed. This section may contain several procedures, used in different stages of an investigation, e.g. for sample preparation, sample testing, post-testing diagnostics, and data analysis. It is useful in such cases to first briefly describe the overall procedure and only then describe the details of the specific procedures.

Typically, some experimental parameters are fixed throughout an investigation, while others are varied in order to determine their effect. Although many of these parameters may have been mentioned in the foregoing apparatus and procedure text, it is good practice to summarize the parameters in a table near the end of this section. First, the fixed parameters should be listed, with their specific values. Then the variables should be listed, indicating either the range over which the variable was varied or the specific values which were employed. An example is shown in Table 2.20. Whenever symbols are used, such as in Table 2.20, the authors must define them at their first appearance. It is critical that the reader know the conditions under which each experimental result was obtained. By presenting such a table, the authors need to refer to it once in order to declare the values of the fixed parameters. In the Results section, the author must be certain to declare the values of all of the variables adjacent to each result.

Table 2.20. Example of a table summarizing fixed parameters and experimental variables.

Fixed Parameters		Value
Cathode diameter		50 mm
Anode i.d.		160 mm
Axial Magnetic Field		100 mT
Variables	**Symbol**	**Value**
Cathode Materials		Zr, Hf, Ti
Cathode Current	I	50–150 A
Deposition Time	T_d	60–180 s

2.6 Results

The Results section is the heart of a research report, and all of the other sections revolve around it: first the Introduction presents the background, then the Methodology section details how the results were obtained, then the Results section presents these results, then the Discussion section interprets and explains the results and their significance, and finally the Conclusions section concisely summarizes what was learned from the results, and explicitly answers the research question.

The Results section answers the basic question "What did I obtain or observe?" A result in a research paper is an observation or measurement that any experienced researcher in the authors' field can exactly reproduce by following the "recipe" given in the preceding Methodology section.

Engineering and science papers usually present the details of their results in figures and tables. These always have a caption so that the figure or table, together with its caption, is sufficiently complete that the reader can understand it without repeatedly referring to the body text. The figures and tables are also always accompanied by body text, and some results are only presented in the body text. The body text describes, summarizes and interprets these figures and tables with sufficient detail so that the reader need not refer repeatedly to the figure or table. Guidelines for figures and tables, including captions, are in section 2.6.1. Guidelines for the body text are in section 2.6.2.

2.6.1 Figures and tables

Ideally, tables and figures should be designed so that their key points are understandable to readers knowledgeable in the field, without reference to the text, even if they do not know the text language. The last requirement may be thought of as the "illiterate man's rule" — i.e. the principle that figures should be understandable even if the reader cannot read the text language.

The conditions by which each figure or table was produced (e.g. experimental conditions, theoretical assumptions, and parameter choices) must be **absolutely clear** for **each** result presented. A fundamental rule is that sufficient detail must be given to allow exact duplication of each and every result. All of the experimental conditions which were constant should be specified in the Methodology section. Alternatively, constant conditions may be detailed in the beginning of the Results section. All variable conditions or parameters must be specified for each and every result, or group of results, either in accompanying text (in the "location sentence" to be described in section 2.6.2) and/or directly in the relevant figure or table (i.e. using the "heads-up display" principle) or its caption.

A "heads-up display" gives the reader as much information as possible in the figure itself, e.g., identifying each curve on a multi-curve graph with a label, and stating the parameter choices for the entire graph by a title within the graph. This alleviates the need for the readers' eyes to jump back and forth between a figure and its legend, caption, or worse, to the body of the text. On the other hand, figures and tables should not be too crowded. There may be a trade-off between making the figures and tables sufficiently detailed and making them readable.

Learn the typographical requirements of the intended journal or graduate degree committee, and adhere to them. Generally, letters and symbols which are too small, and lines which are too thin, will not be legible, especially if the material is photographically reduced to a smaller size before publication.

Carefully consider the best method to present each result. First, determine the most important aspect of the result, and then design the presentation to emphasize and clearly present this aspect. One common problem is the tendency to base figures on readily available material, e.g.

oscilloscope recordings, standard output from various diagnostic instruments such as XPS or XRD spectra, readily available graph formats from Excel or some other software package, etc. While these tools may be useful, their availability is no excuse for not using the most appropriate format, even if so doing requires considerably more effort.

2.6.1.1 Numerical data – graph or table?

The most common forms of results are various types of numerical data. The first question to decide is whether to present such data in a table or in a graph. If the most important aspect of the data is the exact numerical value, use a table. If the most important aspect is comparison, dependence, variation or trend, use a graph.

2.6.1.2 Tables

Use tables only when the most important aspect of the data is the **numerical values**. Think carefully about the number of significant figures that are displayed — inexperienced authors tend to present longer numbers than are justified. Numerical values should be rounded to the number of significant figures justified by the accuracy of the measurement. The accuracy should be estimated, explained and indicated by tolerance values (e.g. 34.541 ± 0.003 g). Units, if appropriate, should be given in the column heading.

Each table, with its caption, should present all of the data needed to convey the authors' point. The table design should not require the reader to switch back and forth between multiple tables to understand some particular technical point. However, the number of rows and columns should be minimized, so as not to overwhelm the reader. Data which are identical in all rows or all columns may be presented in the text or in the caption, and that row or column eliminated. An example is presented in Table 2.21.

Table 2.21. Good practice table example

Caption, sufficiently complete so table can be understood without repeated reference to the body text

Constant parameters given in caption, not in table column

Table XX. Parameters A and b used to fit experimental erosion rate to theoretical curves in the form of $G_a/G_c = A\,(q_a/q_c)^{-b}$ and the resultant coefficient of determination, R^2. The cathode was graphite, and the processed fluid volume was 300 ml in all cases.

Anode	0.1% H$_2$O$_2$ added?	A (±0.002)	b (±0.001)	R^2
Ti	Yes	0.398	1.511	0.87
Fe	Yes	0.954	0.936	0.89
Cu	Yes	2.822	1.039	0.92
Ti	No	0.402	1.482	0.88
Fe	No	0.946	0.962	0.98
Cu	No	2.496	0.956	0.94

Accuracy indicated

Data aligned at decimal point
Number of significant figures does not exceed accuracy

2.6.1.3 Graphs

A graph should be used where **dependence, variation or comparison** is the most important aspect of the data. When the measured quantity is presented for different cases or materials, i.e. as a function of a non-quantifiable variable, the bar graph is most appropriate. When a measured quantity is a function of some quantifiable variable, usually a line graph (called a "scatter graph" in Excel) is most appropriate.

When a measured quantity depends on several variables, the graph should be designed so that the most important dependencies are presented in the form of y(x) curves, where y is the measured quantity, and x is the most important variable. The dependence on other variables can be presented by displaying several curves, each with different values of the less important variables. However, if the dependence on some variable is important, the reader should **not** be expected to **derive** the dependence by comparison of various curves. Where dependence on multiple variables is important, several graphs or three-dimensional graphs may be needed to present these dependencies clearly to the reader.

Clearly label all of graph axes and explicitly state the units. Where possible, present the absolute values of measured quantities, noting the conditions by which they were produced, rather than arbitrary units, even when the absolute value is not needed to make the authors' point in the present paper. The absolute values may be valuable to future researchers for purposes not evident to the present authors.

When stating units, the form ×1000 should never be used, as it is readily misunderstood (about half the readers will understand this to mean that the author has multiplied each value by 1000, so the reader must divide by 1000 to get the correct value, and the other half will understand that the reader must multiply by 1000). Either the values on the scale should be written fully, or accepted prefixes should be used (e.g. instead of (V, ×1000), use mV or kV as appropriate).

Typographically, it is preferable for the graphs to have a white background. Grid lines should be used only if obtaining numerical values from the graphs might seem useful. In that case, only major grids should be used, and both horizontal and vertical grid lines should be displayed. Minor grid lines are unduly distracting. If accurate reading of numerical values is needed, an alternative presentation should be considered. A complete listing of experimental values might be included in an online data base, or numerical fitting parameters might be presented in a table. Annotated examples of good practice use of line graphs follow in Figure 2.4 and Figure 2.5, together with the accompanying body text, which will be explained in section 2.6.2.

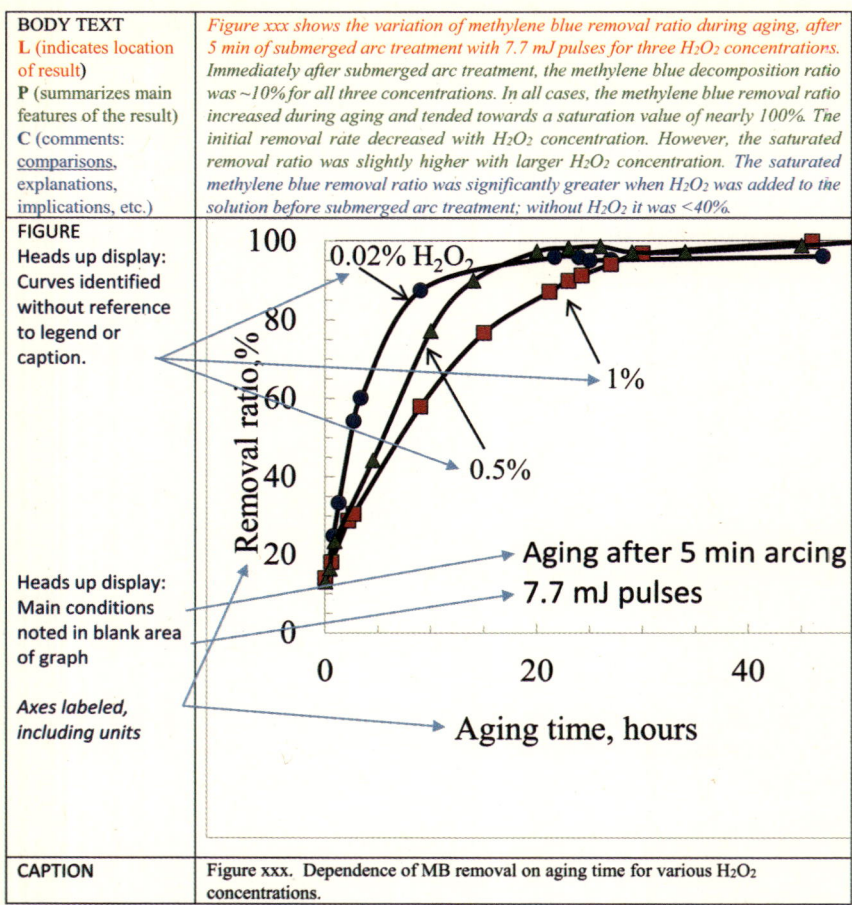

Figure 2.4. Good practice example of a line graph. L, P, and C sentences are explained in section 2.6.2.

BODY TEXT **L** (indicates location of result) **P** (summarizes main features of the result) **C** (comments: comparisons, explanations, implications, etc.	*Figure yyy shows the influence of H_2O_2 concentration on the initial methylene blue removal rate during aging after 5 min of submerged arc treatment with 7.7 mJ pulses. It is derived from Fig. xxx. The initial removal rate was highest for low H_2O_2 concentration, e.g. ~17 %/hr. The rate decreased with H_2O_2 concentration, steeply at first, and then gradually. Although the initial removal rate decreased with H_2O_2 concentration, the final removal ratio increased with H_2O_2 concentration, as shown in Fig. xxx.*
FIGURE Axes labeled, including units	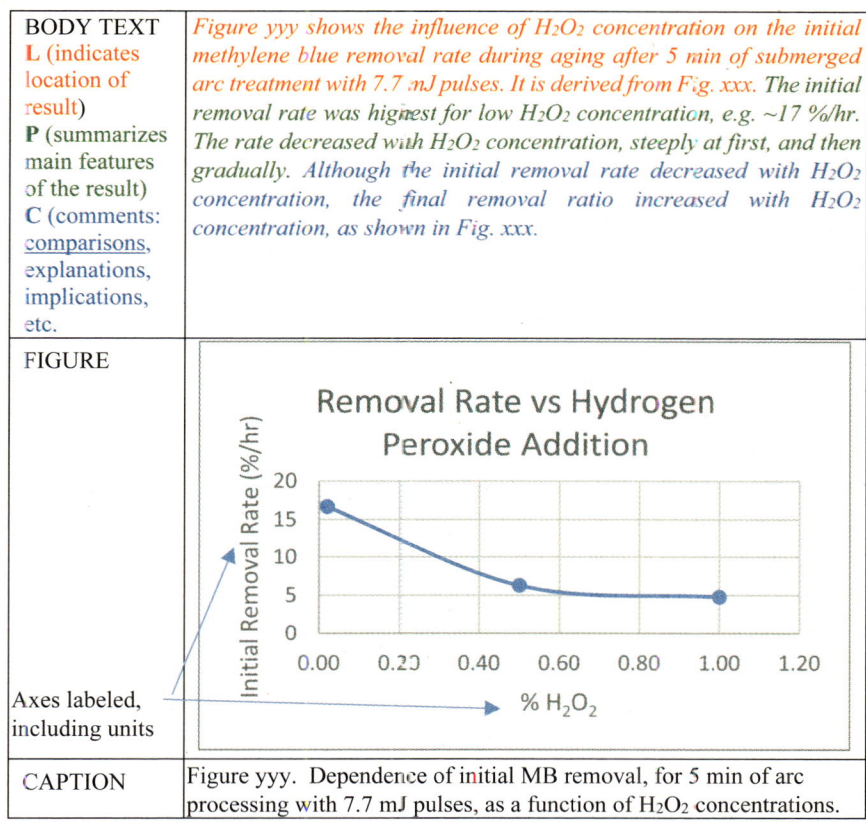
CAPTION	Figure yyy. Dependence of initial MB removal, for 5 min of arc processing with 7.7 mJ pulses, as a function of H_2O_2 concentrations.

Figure 2.5. Good practice graph example, illustrating a method of showing the trend of variation with a parameter from the example in Figure 2.4.

2.6.1.4 *Photographs and micrographs*

Some results are the appearance of specimens, e.g. before and after some process or procedure, and are often best conveyed as photographs or micrographs. The scale of all such photographs must be indicated by a scale bar in micrographs, and either a scale bar or inclusion of some object of known size in normal photographs. The authors, through months of study, readily identify important features in such photographs, but most readers cannot. Therefore, it is imperative that key features be identified with arrows and labels. An annotated example of good practice presentation of a micrograph follows in Figure 2.6.

BODY TEXT **L** (indicates location of result) **P** (summarizes main features of the result) **C** (implication)	*An SEM micrograph within the region of a Ni-coated glass substrate affected by a single 20 µs pulse with a peak current of 20 A between the Ni coating and a graphite electrode in atmospheric air is shown in Fig. zzz. A forest of erect multi-wall carbon nanotubes (CNTs) with lengths in the range of 0.2–1 µm was observed. One end of each CNT was attached to the Ni film, while the top end was free. Approximately 250 000 CNTs were produced in the affected region, which had a width of 276 µm. Assuming that the CNTs formed during the pulse, a CNT growth rate of up to 5 cm/s is inferred.*
FIGURE	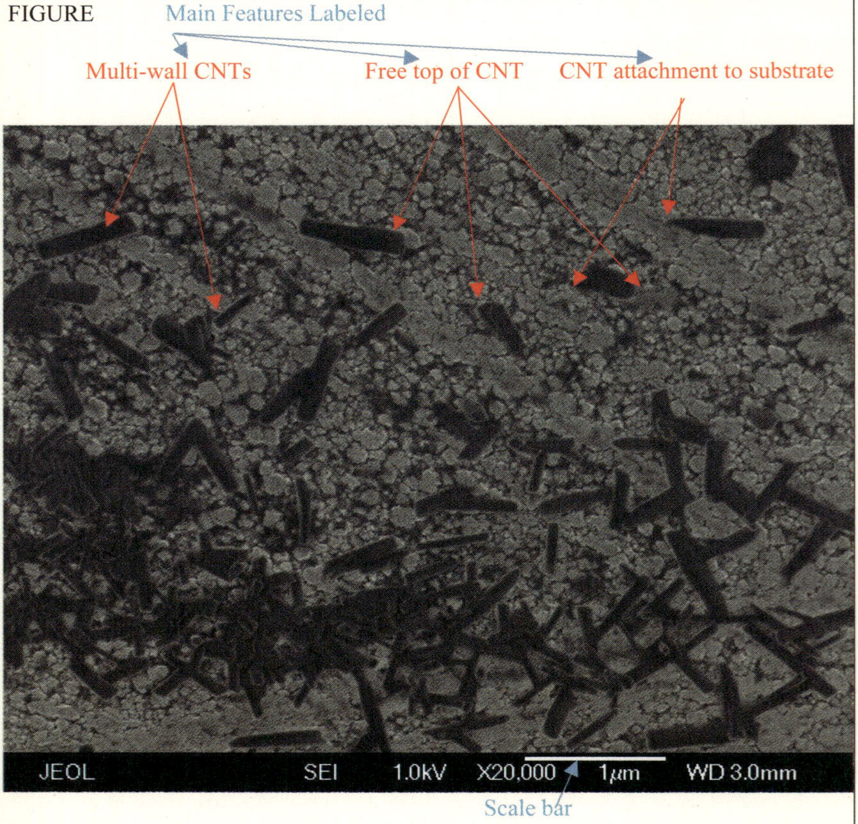 Main Features Labeled: Multi-wall CNTs, Free top of CNT, CNT attachment to substrate. Scale bar.
CAPTION	Figure zzz. Scanning electron micrograph of carbon nano-tubes produced by a single 20 A peak current pulse of 20 µs duration on a Ni-coated glass substrate.

Figure 2.6 Best practice example of a photograph or micrograph. L, P, and C are explained in section 2.6.2.

2.6.1.5 Captions

All figures and tables must have a caption. The caption starts with the title of the figure or table. Then sufficient information should be given so that the figure or table can be understood in the general context of the paper without referring to the body text.

It is good practice to use the automatic numbering feature of most word processors to assign figure and table numbers. References to the figures and tables, e.g. in the location sentence of the text (see section 2.6.2), should use the cross-reference feature. Using these features allows inserting, rearranging, and deleting figures without needing to manually change the numbering throughout the text.

2.6.2 Text

The Results section contains body text which is keyed to the figures and tables. The underlying principle is that the text should present the reader with a complete description of the results that can be understood without actually looking at the figures and tables. This principle might be entitled "blind man's rule" — a blind person having this text read to him should be able to understand all of the results. If this rule is followed by the authors, sighted readers can look at the figures and tables whenever they please, i.e. before reading the text, in the middle of reading the text, whenever a figure is mentioned, or after reading the text. By preparing the text in this manner, the writer helps the reader best absorb the material, since the reader can choose the reading style most suitable for him.

Three types of sentences are used in the text:
- *Location* (**L**) sentences indicate which figure or table contains a particular result.
- *Presentation* (**P**) sentences present the most important findings.
- *Comment* (**C**) sentences comment on and interpret the results.

A typical **L** sentence might be *Figure 5 presents the voltage waveform for a SiC sample measured at a temperature of 27 °C*. The **L** sentence is a good place to indicate the value of variable parameters (e.g. SiC sample, 27 °C temperature), so that all of the conditions required to obtain the result are given to the reader. Generally, **L** sentences are written in the present tense. Either active or passive voice is appropriate.

P sentences summarize the most important aspects of the result. A scientific result has almost "holy" status, and must satisfy the reproducibility criterion, i.e. a skilled researcher elsewhere, who follows the recipe in the procedure section, must obtain exactly the same result, within experimental error. **P** sentences are used to report these "holy" results, and must never be "contaminated" by statements having a lesser status, e.g. interpretations or explanations. While all researchers performing the same experiment or mathematical analysis must reach the same result, their interpretation or explanation of the result may legitimately differ. For this reason, it is the convention that interpretation and explanation are completely separated from results, both in terms of location (by placing them in separate "comment" sentences or in a separate Discussion section) and by linguistic style (results are expressed decisively while interpretations are expressed modestly and tentatively).

Some results are presented only in **P** sentences, i.e. without accompanying figures or tables. For example, *the sample had a pinkish matte appearance and it emitted an acrid aroma.*

P sentences accompanying images of various types should first describe the usual, ordinary, and expected features observed in the image, without interpretation. In other words, describe exactly what you see. Then describe unusual or special features. All of the described features should be labeled in the figure.

P sentences accompanying graphs should describe the dependence seen in the figure, again without any interpretation or explanation. The description should be such that the reader will be able to visualize the graph without looking at it. The description should detail precisely what is seen in the graph. In the case of line graphs having curves with different parameter values, first the $y(x)$ dependencies of each curve should be described and then the various curves can be compared.

The **P** sentence should be written in the past tense to report experimental results, or other results which depend on specific conditions. The present tense may be used, however, to report a result which is always applicable, i.e. not dependent on specific conditions.

The **P** sentence should be as precise as needed by the subsequent discussion of the results. Often being precise requires only a little effort and forethought, and not much space. Consider the following four sentences, all describing the same $y(x)$ curve:

(i)   It may be seen that y depends on x.
(ii) It may be seen that y increases with x.
(iii) It may be seen that y increases linearly with x.
(iv) It may be seen that $y \approx 2.3 x + 32$.

Sentence *(i)* conveys very little information. Most likely, if y did not depend on x, the curve would not be presented. Sentence *(ii)* conveys far more information than sentence *(i)*, yet it contains exactly the same number of words. Sentence *(iii)* conveys even more information, but it is only a little longer than sentence *(ii)*. Finally, sentence *(iv)* is the most precise, and it is also the shortest. However, the reader should not be burdened with information this precise unless these numerical values are important and referred to later in the paper.

Comment (**C**) sentences may be used to compare, interpret or explain a result which was just presented in a **P** sentence. Use present tense for comparisons, modal auxiliaries (e.g. *may, can*) for possible explanations and tentative verbs (e.g. *is likely, appears that*) for generalizations. Some examples of L, P, and C sentences are given in Table 2.22.

Table 2.22.   Sentence examples in the Results section.
L - location sentence, **P** - presentation sentence, **C** - comment sentence.

L	*The correlation parameters as a function of distance from the jet outlet are shown in Fig. 3.*
P	*It may be seen that the correlation decreased steeply with distance, and became negligible after 5 cm.*
C	*These results differ significantly from those observed with conventional jets.*
L&P	*The wavelet intensity had a Gaussian temporal profile, whose width decreased with the distance between the sources, as may be seen in Fig. 4.*
C	*This is similar to the results from ring sources.*

Note that **L** and **P** sentences may be combined, but **P** and **C** sentences should **never** be combined. **C** sentences should be brief, pertain to the immediate result, and convey information that the reader can best absorb at this point in the paper, i.e. immediately following the result. Comments relating to several results or which are long and involved are best incorporated into a separate Discussion section.

Although many papers are published with a combined "Results and Discussion" section, this practice should be avoided to prevent confusing "results" with interpretations and explanations, and so that all of the results are presented to the reader before their interpretation and explanation. For further detail, see the text box in section 2.3.

Annotated best-practice figures with their body text are presented in Figures 2.4–2.6 above. Further examples of figures together with their accompanying body text are given in Table 2.23.

Table 2.23. Examples of figures and text in the Results section.
L - location sentence, P - presentation sentence, C - comment sentence.

Text:	Figure:
L: *The dielectric loss factors of pure epoxy resin and epoxy/crepe paper composites at 160°C are shown in Figure 1.* **P**: *For the ER sample, ε'' initially decreased with f until 20 Hz, and then increased forming a broad strong peak centered at 200 Hz, and then decreased at higher frequencies. In contrast, ε'' of the composite samples decreased monotonically with f, and was identical to ε'' of the epoxy sample for f > 200 Hz. At low frequencies, the samples containing more crepe had higher values of ε''.* **C**: *This behavior is attributed to higher moisture content in the samples containing more paper.*	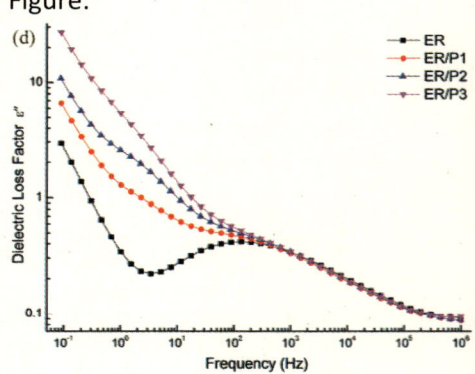 Figure 1. Dielectric loss factors of pure epoxy resin and epoxy/crepe paper composites at 160°C.

Text:	Figure:
L: Figure 2 shows the dependence of the conductivity of Al on the density at 10 000 K. **P:** The calculated conductivity increased only slightly with density below 10^{-3} g/cm^{-3} but increased sharply after that. **C:** The sharp increase in conductivity was due to the effect of pressure ionization and the model described this effect accurately.	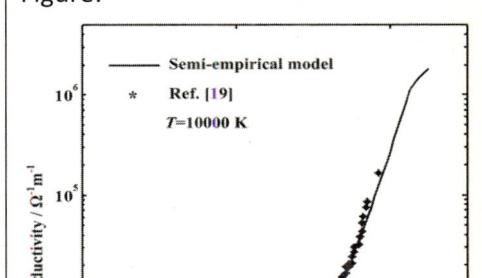 Figure 2. Conductivity of Al as a function of density. T = 10 000 K.
L: The phase curves of the analytical model and the real system are shown in Figure 3. **P:** It may be seen that the phase lag of both curves increased with frequency. The phase lag of the real system is larger than that of the analytical model, and the phase discrepancy between the two curves increases with frequency. **C:** The phase error between the real system and the analytical model is supposed to be caused by eddy currents in the solid-core construction.	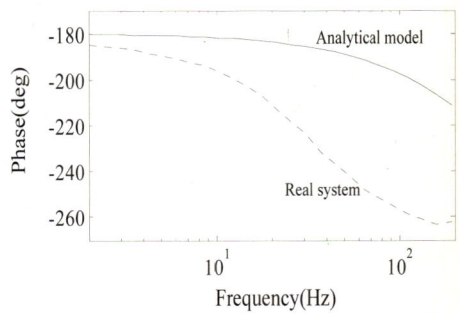 Figure 3. Phase curves of the analytical model and the real system.

Text:	Figure:
L: The waveforms of discharge current and EMI induced voltage are shown in Figure 4. *P: It can be seen that the discharge current was a damped sinusoid, with a period of about 30 μs and a negative peak of about 46 kA, and its corresponding EMI induced voltage was a pulsed voltage with a peak of 2.61 V.* *C: The results in Figure 4 are consistent with the theoretical analysis and the simulation results.*	 Figure 4. Discharge current and EMI induced voltage waveforms.
Text: *L: The spectrum generated by a simple periodic refractive index variation is shown in Figure 5.* *P: Suppression occurs at a central wavelength of 1550.0 nm, with 30% of the input power transmitted through the grating.* *C: This simple grating design suppresses a range of wavelengths on either side of the Bragg value.*	**Figure:** 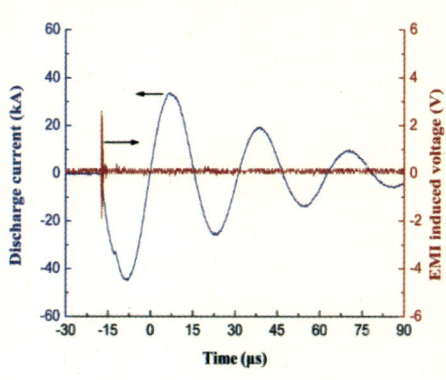 Figure 5. Transmission spectrum generated by a refractive index which varies periodically.

Text:

L: X-ray diffraction (XRD) patterns of the (002) ZnO peak for films with thicknesses of 40, 80 and 180 nm are shown in figure 6.
P: The XRD diffraction peak positions of the films shifted closer to the reference peak at 40.2° for increasing film thickness.
C: The shift in the peak position indicated a reduction in lattice strain as the ZnO films increased in thickness.

Figure:

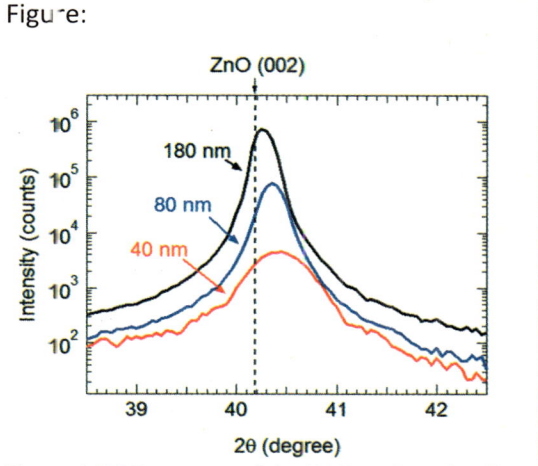

Figure 6. XRD patterns of the (002) ZnO peak. Films had thickness of 40, 80, and 180 nm.

2.7 Theoretical Papers

The overall organization of theoretical papers is similar to experimental papers, as illustrated in Table 2.24. Both experimental and theoretical papers have an Abstract, Introduction, body, Discussion, and Conclusions. The difference is in the body — the guidelines for all of the other sections are identical.

In both theoretical and experimental papers, the body answers the questions: What did you do? What did you get? Although the body of the experimental research report contains sections entitled Methodology and Results or something similar, there is more variation among theoretical papers.

Two prototype body organizations will be examined in this section: "Physical Models" and "Mathematical Proofs". A comparison of the organization of the body for these cases, as well as experimental papers, is summarized in Table 2.25.

Table 2.24. Overall organization of research reports.

Section	Content/Answered Question
Abstract	Short standalone and informative summary.
1 Introduction	What are we talking about?
2 Body	What did we do? What did we get?
3 Discussion	So what?
4 Conclusions	Answer research question. ~3 key points we want the reader to remember.

Table 2.25. Organization of the body of experimental and two types of theoretical research reports.

	Experimental Paper	Physical Model	Mathematical Proof
Body	Methodology	Model Assumptions	Explanation and description of the hypothesis
		Derivation of Equations	Formal statement of the hypothesis
		Solution Method	
	Results	Results	Proof

2.7.1 Physical models

A physical model is a simplified theoretical description of a phenomenon or scenario. Generally, a model starts with well-established and accepted physical laws expressed by equations, and then certain phenomena expressed as terms in these equations are retained, while others are neglected. Physical models are frequently studied because solving a full set of equations is often neither practical nor useful. A full solution will not provide physical insight into what factors are important and which are not. Models, on the other hand, are more readily solved, and the process of choosing, and later verifying, what to take into account and what to neglect, is insightful.

In both experimental and model papers, evidence is presented neutrally first, and then conclusions are drawn. The model body is typically organized in four sections:
- Description of the Model or Model Assumptions
- Derivation of Equations
- Solution Method
- Calculation of Results

2.7.1.1 Model description

The Model Description section typically starts with a description of the physical processes, based on previous theoretical and/or experimental work. It continues with preliminary estimates of the processes or terms, in order to choose the most significant. It concludes with a clear statement of the model assumptions, i.e. which processes, factors, and terms are included in the model, and which are neglected.

2.7.1.2 Equation derivation

The Equation Derivation section first states the initial equations, with references, and clearly states how to proceed from one equation to the next. It defines all symbols, whether they are conventionally used in the literature or not, and all subscripts and superscripts. The explanation of the development of the equations should be understandable to beginning graduate students in the field. This will be in the form of a verbal description, indicating how the preceding equations were manipulated to arrive at the next equation. Sufficient detail must be provided so that results can be duplicated.

2.7.1.3 Solution method

The Solution Method section details how the final set of equations produced in the Equation Derivation section is solved. In some cases, this may be unnecessary if, for example, an analytic solution is produced in that section, or if the solution is so trivial that it can be cursorily explained together with the next section, Results. However, frequently the final set of equations is solved numerically. In that case, the numerical solution method should be described in sufficient detail to allow exact duplication of the results. If a commercially or publicly available software package or a well-known numerical method was used, it should be specified, the specific features or routines used detailed, and a reference provided. However, if a novel method was used, then the method should be described in much greater detail. The distinction here is similar to that between standard and novel apparatus, as presented in section 2.5.2.

2.7.1.4 Results

The Results section presents the results. The input parameters or conditions for each result must be specified. All of the guidelines for experimental results apply here as well. The results will largely be presented in the form of tables and graphs, and the text should use location (**L**), presentation (**P**) and comment (**C**) sentences to describe these results, as described earlier in section 2.6.2. The "heads up display", "blind man's rule", and "illiterate man's rule" as discussed in the experimental results section (section 2.6) apply here, as well as the other guidelines pertaining to the design of the graphs. An example of a results figure and its accompanying text for a physical model paper is presented in Table 2.26.

Table 2.26. Example of a figure and accompanying text for a physical model paper.

Text:	Figure:
L: *Time-averaged ionization rates in the bulk and sheath plasmas as a function of gas pressures are shown in Fig. X for two types of boundary conditions: complete absorption ($r_{L,R} = \gamma_{L,R} = 0$) and partial electron reflection and SEE ($r_{L,R} = \gamma_{L,R} = 0.3$). The numerical parameters are the same as used in Fig. Y.*	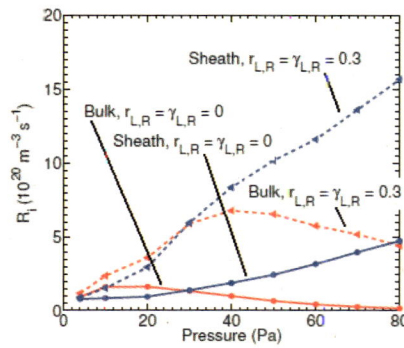
P: *With both boundary conditions, the ionization rates in the sheath increase monotonically with gas pressure (two blue curves), but they increase initially and then decrease at certain pressures in the bulk plasma (two red curves). Above p = 30 Pa, the ionization rates in the sheath become larger than in the bulk. The ionization rates in the sheath and bulk with partial electron reflection and SEE were significantly larger than that with the complete absorption boundary condition.* →	Figure X. Time-averaged ionization rates in bulk and sheath plasmas as a function of the gas pressure. Note that the ionization rate in bulk plasma is defined in the center of the domain, while the ionization rate in the sheath is defined at the border between the sheath and the bulk plasma. Simulation conditions are the same as Fig. xx.
	C: *The larger ionization rate with partial electron reflection and SEE boundary conditions is due to the increasing of the plasma density and particle numbers (see Fig. Z) in the discharge domain. The ionization in the sheath becoming dominant at pressures larger than 30 Pa can be explained by the transition from the 'alpha' to the 'gamma' mode of the CCRF discharge, while most of the ionization collisions occur close to the sheath-bulk border in a 'gamma' mode discharge.*

2.7.2 Mathematical proofs

In mathematical proof papers, a hypothesis (e.g. theorem, lemma, etc.) is stated first and then proven. The body is typically organized in three sections:

(1) Explanation and description of the hypothesis. All terms and symbols are defined.
(2) Brief formal statement of the hypothesis (theorem, lemma, etc.). The statement is normally in the form of a one-line equation or mathematical statement.
(3) Proof — sufficiently detailed so it can be reproduced elsewhere

Appendices are often used in both types of theoretical papers to shorten the body and make it more readable. In physical models, equation derivations are often placed in an appendix. In mathematical proofs, detailed proofs are commonly located in an appendix. Because most readers have faith in the authors and will skip the appendices, the body should be written so that it can be read without <u>ever</u> referring to the appendix. All symbols used in the body should be defined in the body. The objective of each appendix, and the results obtained in the appendix that are used in the body of the paper, should be stated clearly in the body of the paper. The appendix should be written such that it can be read without referring constantly to the body. Start each appendix with a brief statement of the objective of the appendix and summarize the principal conclusion at its end.

2.7.3 Nomenclature and symbols

The most common problem in theoretical papers is undefined and inconsistently used symbols. It is best to minimize the number of symbols. Choose symbols commonly used in English scientific literature, or define symbols using initial letters based on their **English** name (and not on their name in your native language).

If the paper has many symbols (>10), it is strongly recommended that the authors prepare a nomenclature table for their own use during writing and editing. This table should have 4 columns, as shown in Table 2.27.

The first two columns contain the symbols (listed in alphabetical order) and their definition. A nomenclature table containing only these two columns can be optionally included in the paper. The third and fourth columns list all the pages where each symbol appears, and the page where the symbol is defined. During proof-reading, the author should check each symbol at each appearance against columns 1–2 to ensure consistent usage and fill in columns 3–4 to verify that each symbol is defined adjacent to the location where the symbol first appears.

Table 2.27. Sample nomenclature table.

Symbol	Definition	Pages	Defined on page
A	Waveguide width	2,3,4	2
B	Waveguide height	2,3,4	2
V	Velocity	3,5	3
Subscripts			
E	Electron	3, 5	3
I	Ion	3, 5	3

2.8 Discussion

The Results section conveys **information** to the reader — i.e. what happened when a particular procedure was executed under specific conditions. The objective of the Discussion section is to convert this **information** to **knowledge**. This is done by placing the results in their scientific context and answering the question: "So what?" Answering this question typically involves explaining and interpreting the results and their significance. The Discussion is usually the most difficult section to prepare. It requires scientific understanding and judgment. Novice scientists are best helped in preparing the Discussion by extensive reading of the literature and by detailed discussion of their work with mentors and colleagues.

This section should discuss results presented earlier in the Results section, or in the previous literature (with a specific reference). It should **not** introduce "new" results or new facts, i.e. "rabbits should not be pulled out of the hat."

The Discussion section typically opens with specific statements reminding the reader of key information presented previously, then broadens its coverage by interpreting the results and concludes quite broadly by discussing the implications of the present work, as illustrated in Table 2.28. Reminders of previous elements are particularly valuable in long reports, and serve to re-focus the reader's attention on the main issues, after being engaged with the minutiae of the results. They can be used to remind the reader about the purpose of the work, and highlight the most important results.

Table 2.28. Discussion elements.

Reminders (narrow focus)	1. Reference to main objective, hypothesis, or research question
	2. Brief review of the most important results
Interpretations (broadened focus)	3. Justifications
	4. Limitations
	5. Comparisons
	6. Validations
	7. Explanations
Implications (further broadened focus)	8. Generalizations
	9. Significance
	10. Recommendations

The bulk of the Discussion typically interprets the results, through justifications, limitations, comparisons, validations, and explanations. Justifications present arguments and analysis which show that the approach and the subsequent results are valid or at least are not invalid. This might include, for example, statistical analysis to show that trends observed are significantly larger than scatter in the data and analyses of experimental error. In theoretical models, demonstrations of self-consistency may be presented.

Limitations explain how the finite capabilities of the apparatus or methods might impact and limit the results. For example, phenomena faster than the response or sampling time of a recording instrument, or objects smaller than the spatial resolution of a microscope, or chemical components whose concentrations are smaller than the sensitivity of the analysis technique, could not and would not be observed, and this must be taken into account when interpreting the results.

Comparisons may be internal or external. Internal comparisons compare results presented in the Results section. The comparison may be of theoretical and experimental results, between experimental results obtained by different methods, e.g. different diagnostic techniques or under different circumstances (e.g. experimental conditions). The researcher may infer the influence of the different circumstances on the results. External comparisons may consider any of the above with results previously presented in the literature, either by the present or other investigators. External comparisons promote synergy within the scientific community, and the lack of external comparison is a frequent cause for papers to be rejected for publication.

In the Discussion section, specific references to the purpose, hypotheses and findings of the current study generally use the simple past tense of the verb. In comparisons, the present tense is conventionally used.

Perhaps the most important discussion element is explanations. A typical discussion statement is structured with a main clause stating the researcher's position, the word *that*, and a noun clause which conveys the explanation, e.g. *Thus, it was proven that the increased electron emission was caused by process sample heating.*

In experimental work, the underlying cause of the experimental results is usually interesting. In cases where the theoretical explanation is relatively simple, the entire theory may be developed and presented within the Discussion. In other cases, where the theoretical explanation is lengthy, either a separate theoretical section may be included in the report, or a separate theoretical report may be prepared. In either case, the result of comparing the experimental and theoretical results should be presented in the Discussion. In other cases, explanations may be available in the literature. In the case where multiple explanations already exist, the Discussion might discuss the relative merits of each. In some cases, where no explanation exists and where a full theoretical development is beyond the scope of the present work, the Discussion might include the author's rough ideas about the explanation.

Validation here refers to comparing theoretical results with experimental results (either obtained as part of the same work, or from the literature). It is expected that a theory should be tested experimentally, and the authors of a theoretical work should compare their results with suitable experiments, if they exist. The lack of such comparison, when comparison is possible, is a frequent cause of paper rejection.

There is no obligation for an experimental paper to provide a full theoretical explanation, nor for a theoretical paper to provide experiments to validate the theory. These tasks may be left for future work by either the present authors or other researchers. However, the Discussion should note the absence of an experiment validating the theory or the lack of a theoretical explanation for the experimental results, as the case may be. Preliminary ideas might be outlined regarding possible theoretical explanations.

It is important that the authors understand and correctly categorize the degree of certainty of any explanation presented, and use language that conveys this degree of certainty to the reader. While it is a goal of the scientific method to prove physical laws, such proofs are very rare, and most research reports provide explanations which fall short of satisfying the requirements for a "proof" i.e. 1) that all possible explanations are known, and 2) that a decisive test is performed in which one and only one of the possible explanations can correctly predict. Sometimes the best that an author can provide is a speculation, while in other cases it is highly likely that the offered explanation is correct. Explanations may fall on a continuous scale from "speculation" to "proven". Benchmarks on this scale and sample sentences are shown in Table 2.29.

Similarly, the verb tense used in explanations indicates degree of certainty. The present tense is used when the findings are felt to be generally applicable and the past tense when they are restricted to this study. A modal auxiliary is used for speculative explanations, as in the first example *may be* in Table 2.29 — see section 10.3.1 for further explanation.

Table 2.29. Sliding scale of certainty for explanations

Category	Criteria	Key Words	Examples
Speculation	An idea or ideas that come to mind	may, possible, conceivably	A *possible* explanation is that the specimen *may* be heated by the process.
Likely	Some evidence supports this idea	suggests, indicates	The increased electrical conductivity *suggests* that the sample is heated by the process.
Very likely	Substantial evidence supports this idea	is consistent, strongly suggest	The increased electrical conductivity as well as the increased radiance, *strongly suggest* that the sample is heated by the process.
Most likely	There is more evidence and/or theoretical support for this idea than any other existing idea	most likely	The *most likely* explanation for the increased electron emission is sample heating by the process.
Proven VERY RARE	All possible explanations are on the table, and a decisive test indicates that this idea and only this idea explains the observation	proven, proves, proof, shown, demonstrated	Thus it was *proven* that the increased electron emission is caused by process sample heating.

The discussion of the implications of the current research includes generalizations and recommendations. A given research project only investigates a finite set of conditions, but often the authors believe that the results presented are generally true for a much broader set of conditions. It is likely that the results also apply to intermediate cases which can be interpolated from the presented results. And it is possible that the results also are applicable to other cases which can be extrapolated from the presented results, either quantitatively or merely qualitatively, because of some similarity or underlying principle. This is the place to express such an opinion, but it should be expressed tentatively.

Often there is significance to the research results beyond the obvious. For example, answering the specific "research question" of the present research might allow addressing new or additional questions, or the research results might be exploited for some practical application, e.g. a consumer product, industrial process, or medical therapy. Such significance should be presented modestly.

Guidelines for Modesty in Scientific Writing

Scientific readers distrust reports which are not objective. Immodest language in a paper suggests a lack of objectivity, and undermines the trust which the reader naturally grants authors. Some suggestions are offered below for maintaining modesty in research reports:

- Avoid using the first person (I/we).
- Have a clear gap sentence in the Introduction, but do not explicitly claim to be the first person to address a problem. The reader will infer this on his own.
- Explain why a topic or result is significant or important, but do not rank its significance or importance Avoid statements such as *this is the most important [or significant] result in the past 20 years*. Let the readers rank the significance or importance on their own.
- When presenting explanations, use appropriate phrasing to demonstrate the degrees of uncertainty (see Table 2.29).
- When presenting extrapolations (i.e. suggesting that a result may be valid for cases which were not explicitly studied), use appropriate tentative language.

Finally, the author may recommend some course of action that is suggested by the present research results. This might be follow-up research to explore further aspects of the problem addressed in the current research, to carry it further, or to address new questions. Such a recommendation may be adopted by either the current author, or by anyone else in the scientific community. Scientific, technical or commercial exploitation of the results may be recommended, e.g. to develop a product, process or therapy. Or the author might recommend adopting some public policy, as a direct result of some specific research, e.g. in the realm of ecology or health. In all cases, this section should be brief and firmly based on the specific results presented in the present report.

2.9 Conclusions

Conclusions briefly **summarize** the most important results and their implications. This section delivers the "bottom line" of the research report. The conclusions might appear as the last paragraph of the Discussion, but generally it is better for them to appear in a separate Conclusions section. The results should have been presented in the Results section and the implications developed in the Discussion section. Thus there should be no new facts presented in the Conclusions section. All of the sentences in the Conclusions should be informative, i.e. provide actual information.

It is useful for the author to choose three statements that the reader should remember. One of them must be the **explicit answer to the research question**. Examples are given in Table 2.30.

Table 2.30. Examples of informative conclusion statements, explicitly answering research questions.

Conclusion Statement	Corresponding Research Question
It was found that adhesion of TiN coatings on steel substrates was improved with the application of negative pulsed biasing.	How does pulsed biasing affect the adhesion of TiN coatings on steel substrates?
Epoxy/crepe paper composites have larger values of real permittivity and imaginary permittivity than pure epoxy, and this is due to higher moisture content absorbed by the crepe paper.	How does the permittivity of epoxy/crepe paper composites compare with pure epoxy? Or What factor is responsible for the difference in permittivity between epoxy/crepe paper composites and pure epoxy?

Conclusion Statement	Corresponding Research Question
Of the three methods compared, the maximum braze strength of 106 MPa was achieved using 97Ag-1Cu-2Zr active brazing filler, in a furnace with resistive heating at a temperature of 990°C for 5 minutes.	Which metal-ceramic brazing method produces the strongest junction between alumina and stainless steel?
The solid solution solubility of Na in SnSe is 0.1%, far above the necessary doping concentration to tune the carrier concentration to the optimum thermoelectric properties	What is the solid solution solubility of Na in SnSe, and how does this doping affect the thermoelectric properties?

The Conclusions is not a summary — that function is fulfilled by the Abstract, which will be described in section 2.10. The Conclusions should not include any sentence that does not express a conclusion. Thus, the following should **not** be included:
- recitation of the research question, hypothesis, objectives, or methods
- new facts
- references (neither internal (e.g. see Fig. 3) nor external (e.g. Smith [3])
- merely indicative sentences (i.e. sentences which only indicate the scope of the work, e.g. *The adhesion of TiN coatings on stainless steel substrates was determined as a function of the bias voltage*).

The Conclusions section should be short. In a journal paper, a single brief paragraph is usually sufficient. In a longer research report, such as a thesis, the Conclusions might be a half a page to two pages long.

2.10 Abstract

The Abstract usually appears at the beginning of a research paper, but it should be written, or at least rewritten, last. Many more people will read the Abstract than will read the paper. Readers of journals will often skim abstracts to select the articles which they will read especially where the full article is behind a paywall. In addition, abstracts are published separately in abstract journals and databases. Accordingly, the Abstract should be prepared with special care.

The Abstract should stand alone with no internal (or external) references. It should summarize the key sections of the research paper — Background, Objective, Methodology, most important Results and Conclusions, in 1–2 sentences for each. The Abstract is not the Introduction. Accordingly, the background in the Abstract should be confined to 1–2 sentences. The Abstract should **inform** the reader of the key results, **not** merely **indicate** the scope of the paper. This is accomplished by using **informative** sentences. Table 2.31 presents examples of indicative and informative sentences. The Abstract may include indicative sentences to indicate the overall scope of the work, but it must include informative sentences that actually present results and conclusions.

Table 2.31. Examples of indicative and informative sentences.

Indicative sentence	Informative sentence (do it this way!)
The voltage as a function of temperature was measured.	It was found that the voltage decreased as a function of the temperature, reaching a saturation value of 30 mV.
The real and imaginary permittivity of epoxy/crepe paper composites was measured.	Epoxy/crepe paper composites have larger values of real permittivity and imaginary permittivity than pure epoxy because of their higher moisture content.
The survival of E. Coli *bacteria* was determined as a function of penicillin dosage and temperature.	At 37 °C, the survival of E. Coli *bacteria* decreased with penicillin dosage, and reached 8-log at 10 mg/l.
Vorticity was measured as a function of the protuberance length and flow velocity.	Vorticity increased linearly with the Reynolds number.
The differential quasi-steady state model was compared to the mass action model.	The differential quasi-steady state model closely matched the mass action model during the transient phases and matched exactly at steady state.
The intermediate mass occurrence rate was compared with the predictions of the direct collapse model, and the result was used to evaluate existing theories for supermassive black hole formation.	The intermediate mass occurrence rate of 5% is consistent with the direct collapse model, providing support for the formation of the supermassive black holes via merger of intermediate mass black holes.

Some media, in particular conference proceedings, impose strict length limits on Abstracts. In such cases, background material may be deleted and/or the methodology and the principal results may be combined in a single sentence. For example: *Using an intercellular probe, it was found that the voltage decreased as a function of the temperature, reaching a saturation value of 30 mV.* Use abbreviations and acronyms in the Abstract **only** if a term is used repeatedly within the Abstract, **and** its use will save considerable space in the Abstract, taking into account that each must be defined the first time it is used.

2.11 Title

The title and a detailed outline should be composed at the beginning of the writing process, but the title should be re-evaluated after the paper is written. The title should accurately express the subject of the new results presented. Often the direction of a paper "shifts" during the writing process, and the final direction should be reflected in the title.

The title should be short — no more than two lines; one line is even better. **No abbreviations** should be used — this is important because of keyword archival searches. A commonly-used abbreviation today may not be recognized 50 years from now. The paper should be written as if it has seminal importance in your field and will still be read and referenced by future generations. Some journals also require a very short 'running title' which appears at the head of each page. In such cases, key words reminding the reader of the main aspect of the title should be included in the running title.

2.12 Authors

The authors are those who substantially contributed to the intellectual content of work, which may include planning the research, executing the research and evaluating and interpreting the results. In the West, the "boss" (e.g. manager, supervisor or thesis advisor) is **not** automatically included as an author. The "boss" is included if and only if he or she participated in the planning, direction, execution, evaluation, and/or interpretation of the research. This of course can be a delicate issue, and it is best to have a clear understanding with him/her before beginning to write.

In most journals, the authors are ordered according to the weight of their intellectual contribution. (A few journals order the authors alphabetically.) The first author is the most substantial contributor and usually writes the paper. There can be a delicate question in hierarchical situations such as teacher/student or manager/researcher. Was the role of the student mostly to carry out the instructions of the teacher (i.e., do the "leg work")? Or was the student the main intellectual contributor, and the teacher only gave occasional advice and guidance? Often there may be progression from the former to the latter as the student advances. But some teachers may tend to insist on being listed as first authors, even when the intellectual contribution of the student was significant. Again, the writer should have a clear agreement with his or her supervisor on this point.

Authors should use consistent spelling and first name or initials in all published papers. This facilitates literature searches. Name changes, due to marriage or other reasons, can make this difficult. Some researchers choose to retain the name under which they started publishing for professional purposes.

Affiliation at the time when the work was done, and the mailing address and e-mail should be included for all authors in the byline. Current affiliation and address, if different, should be indicated in a footnote. The exact style may vary according to the journal, hence most provide templates or style guides.

It is professional courtesy towards co-authors to give them the opportunity to review and correct the paper before its submission, and to share with them all correspondence, in particular the reviewers' comments. Likewise, their comments should be solicited on any subsequent re-submission.

The list of authors does not include those whose contribution was purely technical — e.g. soldering wires together, constructing equipment according to specific directions or performing standard tests. Their contribution should be noted in an Acknowledgements paragraph, usually placed at the end of the paper. Likewise, it is appropriate in the Acknowledgements to thank or acknowledge others who assisted or supported the research, including technicians, sources of funding, lenders of equipment or facilities, providers of other services and casual advice givers.

2.13 Abbreviated Research Reports

Abbreviated research reports present new scientific findings to the scientific community without necessarily meeting fully the requirements to place the work in context and providing sufficient detail for duplication. Examples include letters and brief communications in journals, extended abstracts, and papers in conference proceedings, where the length is often strictly limited. The overall objective of the abbreviated research report is the same as the full research report, namely to share scientific findings with the scientific community. However, because of the limitation of space, often only a broad outline of the method is presented and thus there may not be sufficient detail to permit duplication. The background and discussion are limited. Letters are intended to rapidly convey very important or very timely new results. They are often accorded an expedited review and editorial process. Letter authors should follow up at a later date with a full paper which gives sufficient detail for duplication.

Letters, brief communications and extended abstracts are not usually divided into sections with section titles, and might not include an abstract. However, the overall organization is the same as in a full research report, i.e. with an introductory paragraph briefly giving the background and objective, and a number of paragraphs as needed (within the space limitations) giving the methodology, results, discussion and conclusions. The author should strive to meet the standards of a full research report, in particular giving details needed for duplication, to the extent that the space limitations permit.

2.14 Review Papers

Review papers survey the literature and critically examine the research on a topic. Good review papers are often highly cited and contribute greatly to the field. They provide the novice with an overview and the expert with insight. Review papers are generally written by experienced experts who have the necessary perspective to objectively and critically review the topic.

Review papers are not research reports and are organized differently. They begin with an Introduction which 1) explains the topic and its importance, 2) includes a type of gap statement that notes the need for a review paper (for example, the topic was never reviewed or significant advances have occurred since the last review) and 3) states the overall scope and objective of the current paper. The body reviews the main accomplishments of the previous research reports on the topic. This review is organized with appropriate section headers generally according to the methods previously suggested in Section 2.4.2 for the literature survey in the Introduction of the research report. A Discussion section provides the true added value of the paper. It critiques the past work, summarizes conclusions gained from comparing previous work, and formulates unanswered questions that should be addressed in the future.

For further reading:

Felice Frankel and Angela H. De Pace, *Visual Strategies: A Practical Guide to Graphics for Scientists and Engineers*, Yale University Press, 2012.

Hilary Glasman-Deal, *Science Research Writing for Non-native Speakers of English*, Imperial College Press, 2010.

Angelika H. Hofmann, *Scientific Writing and Communication: Papers, Proposals and Presentations*, Second edition, Oxford University Press, 2013.

Robert Weissburg and Suzanne Burker, *Writing Up Research: Experimental Research Report Writing for Students of English*, Prentice Hall Regents, 1990.

Chapter 3

Submitting a Paper and the Review Process

3.1 Before Submitting a Paper: Ethical Issues

3.1.1 *Scientific integrity*

Accurate reporting of what was done and observed in a research project is fundamental. A research report that presents invented results or that omits inconvenient results is unacceptable.

Sometimes research is conducted with an experimental system which is well understood, well controlled, and which gives experimental results that are repeatable to within an insignificant scatter. But frequently this is not the case, particularly at the forefront of science. Sometimes conducting sufficient repeat experiments to establish statistical significance is impossible or impractical. Often, whether to report data with a high degree of scatter, and if so, how, is an issue. Results, even if imperfect, should be reported if they are apt to be useful to the wider scientific community. But the results must be reported honestly — all of the results observed under a particular set of conditions should be reported. Data points should not be eliminated merely because they are outliers or don't fit some particular theoretical model. On the other hand, data obtained under conditions which differed from other data (e.g. the test chamber door was left open by mistake) may either be eliminated or reported together with a description of the exceptional conditions.

Both curiosity and skepticism are useful characteristics for researchers. Curiosity drives the researchers to investigate unusual results in depth, and to try to understand them. Sometimes unusual results and keen observation leads to significant discoveries. But most often unusual and unexpected results, especially those in defiance of well-established scientific laws, stem from experimental mistakes. The researcher's skepticism should be exploited to prevent jumping to conclusions prematurely.

3.1.2 *No plagiarism*

Plagiarism is presenting someone else's results, ideas, or words as your own. It is a "cardinal sin" in the scientific community. Everything in a research report must either be the result of the authors' work or explicitly attributed to the original author by a citation. If a verbatim quote is used (rare in engineering and science), it should either be enclosed in quotation marks ("…") or indented, and the original work must be cited.

3.1.3 *No copyright violations*

Material published in books and journals, and often other written material, is protected by a copyright. The copyright is a form of intellectual property (IP), as are patents and trademarks, that gives its holder the exclusive right to copy the material. Researchers may, however, make "fair use" of published material, which may include summarizing a published work in the literature review, provided the original paper is cited. A short quotation may be included with a citation. However, including previously published figures and tables, even if cited, generally requires permission of the copyright holder, i.e. the journal or book publisher. Usually such permission is granted, and journals often provide a form through which to request this permission from other journals.

3.1.4 *No double publication*

The same material may not be published twice nor should it be submitted simultaneously to more than one publication. Publications in this context include books and journals.

Conference proceedings can be a grey area — they are not always formally published. For example, if the conference proceedings are distributed only to attendees and are not generally available to others, they might not be considered a publication. Even if the conference proceedings were published, the material might be used as a basis for a longer and more detailed article in an archival journal.

A series of articles relating to ongoing research at different stages may have some minor overlap for completeness and clarity. However, it should be clear, via reference, what is old and what is new.

3.2 The Review and Publication Process

Authors submit their paper to the editor of a journal. The editor will then send the paper for peer review. The term peer means equal, and implies other working scientists rather than some special panel of experts. However, reviewers selected are typically knowledgeable in the field and their opinion is valued by the editor. The reviewers read the paper and file a report which contains a recommendation. If the recommendation is not to publish, the authors are so informed. If the paper is conditionally accepted, the authors will be requested to revise their paper and resubmit.

In traditional publishing, authors are required to sign a form transferring the copyright to the publisher of the journal and in some journals to arrange payment of page charges. However, there is an increasing tendency towards open access publication. In open access publication, the legal rights of authors and users are specified in a Creative Commons agreement. The authors pay an open access fee and usually retain ownership of the copyright.

If the paper is accepted, it will be processed by the journal for publication. Traditionally this includes copy editing, to ensure meeting the journal's style standards, and typesetting. However, with electronic publication this burden is increasingly imposed on the authors. After the paper is typeset, a "page proof" (sometimes called a "galley proof") is sent to the authors, who must quickly and thoroughly check and if needed correct it and return it to the journal, which will then generally publish the paper in the next available issue. In the following paragraphs, the various stages will be elaborated in more detail, and the manner in which the authors should interact with the process will be explained.

Although all of the authors of a multi-author paper are collectively responsible for its content, one author handles the mechanics of interacting with the publication process. Usually this is the first author. Sometimes another "corresponding author" is designated, e.g. when the first author's address is expected to change during the publication process.

3.2.1 *Submission*

Journals commonly have an online mechanism for submitting papers. Many journals provide a WORD or LaTeX template via their website. In the absence of an online submission mechanism, the submission directions provided by the journal, typically found on their website or on the inner cover of an issue of the journal, will serve as a guide.

The paper manuscript should be accompanied by a cover letter. This is a form of a business letter (see section 9.1), which generally has a very short body (e.g. 1-sentence — *Please find enclosed a paper entitled XXXX by Y and Z, submitted for your consideration for publication in the Journal of JJJJ*). The cover letter may be slightly longer if there is something special about the submission, e.g. if the article is part of a series. The letter should be written on the author's institutional letterhead and include all of the relevant communications data (mailing address, e-mail address, and phone numbers).

3.2.2 *Reviewer selection*

Editors will generally seek reviewers who have expertise in the particular sub-area of the paper, as determined by looking at the title, abstract and keywords. Thus it is important for the authors to ensure that these elements of the paper are as accurate as possible, and use the terminology common in the field.

Editors will frequently ask authors who have submitted good papers to the same journal to review some other paper. This may happen at a relatively early stage of a researcher's career, even while still a student, if the paper was particularly good. Guidelines for reviewing a paper are provided in section 3.3. Some journals solicit suggestions for reviewers from the authors, e.g. via their online submission system. Authors should suggest reviewers, if requested, for this can shorten the review process. The suggested reviewers should not be from the same institution as the authors, and if possible should not be closely connected with them, e.g. by having an ongoing collaboration. The number of reviewers varies among the journals. Traditionally, the identity of the reviewers is confidential, in order to encourage a frank evaluation.

3.2.3 *Editorial decision and the author's response*

Generally, the editor follows the recommendation of the reviewers. If their recommendations disagree, the editor may follow one of several alternatives.
(1) Very selective journals might opt to follow the harshest recommendation.
(2) If there were more 3 or more reviewers, the majority or consensus opinion might be adopted.
(3) If there were two diverging reviews, the editor might seek the opinion of a third reviewer. This might be a request to simply review the paper. Alternatively, the editor might seek the opinion of someone particularly senior in the field, to adjudicate between the reviewers, in which case the adjudicator will be sent both the submitted paper and the previous reviews.

If the decision is to publish "as is", the author can simply attend to the technical aspects of the publication process (copyright forms, galley proofs, etc.) when they are sent by the journal. This outcome is relatively rare, and most decisions involve a request or demand for revisions, or even rejection. Generally, the editor will inform the author of his decision, enclose the reviews, and possibly may send additional editorial comments.

3.2.3.1 *Acceptance with revisions*

In the case of acceptance with revisions (whether mandatory or suggested), the authors should first carefully read all of the material and consider the reviewers' and editor's comments. The authors should give *prima facie* credit to the reviewers that they are experts in the field and experienced readers. If they claim that something is wrong in the paper, they are probably correct. And if it appears that they didn't understand some point, it is most likely that that authors did not present it clearly and that many other readers will likewise have difficulty in understanding it. Thus, it is worthwhile to comply with all of the reviewers' requests whether mandatory or merely suggested. Doing so will most likely produce a better paper.

However, reviewers are human beings, and are sometimes overworked, lazy, not so expert in the field, or self-serving. If the authors cannot in good conscience comply with a requested or even a required revision, they should rebut that point. The rebuttal should be polite and factual, dealing with the scientific issue, and not the reviewer.

The author's response includes the following elements:
(1) The revised manuscript, preferably with the revisions highlighted. It is good practice also to submit a clean copy of the revised manuscript.
(2) A cover letter (sometimes called a response letter) addressed to the editor, indicating:
   (a) the location of each revision, and
   (b) rebuttal of any of the reviewers' comments with which the authors could not comply.

The locations and rebuttals should be numbered and keyed to the reviewers' numbering of their comments. If the reviewers did not number their comments, the author may manually number them. If the number of comments is sufficiently small so that the author's response letter is less than two pages, the response may be included in the body of the cover letter. If the letter with comments is longer than two pages, the comments should be included in a table appended to the cover letter.

It should be emphasized that the authors' response is essentially the revised manuscript and the revisions therein. The cover letter or appended table should not contain an explanation of the revisions, but only their location. In other words, the revisions must speak for themselves and be sufficiently clear so that the editor and reviewers understand them. The reason for this is that the ordinary readers of the journal will not have the benefit of any supplementary explanations. If the revised text is not clear to the reviewer, it will not be clear to the ordinary reader either.

All of the author's correspondence is with the editor. The author should not address the reviewers at any point in the process. If a review was particularly helpful in improving the paper, the author may express their appreciation of the review to the editor.

3.2.3.2 *Rejection*

The authors have several options if their paper is rejected:
(1) If the authors are convinced by the reviews that their paper is unworthy of publication in any form, they may accept the rejection, and no response from them is needed.
(2) If the reviewers recommend rejecting the paper because of a correctable defect, the authors may correct the defect and resubmit the paper, i.e. respond just as if the paper was accepted with mandatory revision. Some journals have a formal appeal process for considering manuscripts revised to answer the reviewers' criticism.
(3) If the authors disagree with the reviewer, they may send the editor a rebuttal. It should be polite and factual, and deal with the scientific issues, and not any personal issues. Some journals have a formal appeal process in which authors can rebut the reviewers.
(4) If the authors are convinced that their paper is worthy of publication despite the reviewers' comments, they may submit their paper to another journal. In this case, it would be wise to revise their paper in light of the referees' comments, i.e. incorporating whatever positive suggestions the reviewers provided, and thus improve the paper. Once the paper has been rejected and is no longer being considered by the first journal, there is no need to inform the second journal that the paper had been previously rejected unless the editor specifically asks.

3.2.4 Editor's response to the revised paper

If the required revisions were relatively minor or simple and if the editor is convinced that the authors complied with the reviewers' comments, then the editor will accept the paper for publication. For this reason, it is important that the authors clearly show the location of the revisions via a table keyed to the reviewers' comments.

If the required revisions were major or complicated, if the editor cannot readily determine if the authors complied with the reviewers' comments, or if the authors ignore or reject the reviewers' comments, the editor will re-send the manuscript with the authors' response to the reviewers or possibly send it to new reviewers.

3.2.5 Technical matters

Once the paper is accepted, the first author or the designated corresponding author will be required to fill in and sign a copyright transfer form or complete the formalities for open access publication. This is essential, as the publisher cannot publish the paper without one or the other.

Some journals require the authors' institution to cover part of the publication costs in the form of "page charges". The authors should be aware of the page charge policy of the journal before they submit. Where there are page charges, a purchase order form is sent to the first or corresponding author after acceptance.

The technical terms "typesetting" and "galley proofs" originate from a bygone era when moveable type was manually set into galleys. Today, all journals accept and process electronic manuscripts. Generally, someone in the editorial office (the copy editor) reviews the manuscript for style and formats the file for printing. When this process is completed, the author is sent a galley proof to expeditiously review, usually within a specified short time frame (i.e. a few days). The main intention of this review is to correct errors that the copy editing and formatting may have introduced, e.g. figures misplaced or oriented incorrectly, typographical errors, etc. The authors may also use this opportunity to correct minor errors of their own, e.g. grammatical, spelling and typographical errors. Substantial corrections are not permitted at this stage. Any major change in the content will require re-submission and a new review.

3.3 Reviewing a Paper

The peer review process fulfills two functions. It preserves the quality standard of the journal. And it provides immediate feedback to the authors on the quality of their work and often gives suggestions for improving their paper.

Almost all journals conduct their review process via a website established by the journal or publisher for this purpose. Typically, the review contains several elements:
(1) Questionnaire. This form forces the reviewer to make discrete choices which describe the state of key editorial elements and, in particular, to give a bottom line recommendation regarding publication (e.g. publish as is, publish with revisions, or reject).
(2) Free-form evaluation.
(3) Marked-up manuscript. Many editors will provide upon request an editable manuscript source file (e.g. in Word *.docx format) on which the reviewer can make minor corrections and insert comments keyed to particular textual passages.
(4) Cover letter. In rare cases where the review procedure is not conducted on-line, the review should be sent to the editor together with the cover letter.

Preparation of each of these elements will be described in the paragraphs below.

3.3.1 *Questionnaire*

The objectives of a questionnaire are to focus the attention of the reviewer on the key issues underlying the decision on the suitability of a paper for publication in a particular journal. The choice of issues is according to the judgment of the editor or whoever prepared the questionnaire. The reviewer is expected to decisively render judgment on each of these issues.

Usually the questionnaire ends with a bottom-line question: whether to accept the paper for publication and under what conditions.

3.3.2 Free-form evaluation

The free-form evaluation allows reviewers to express their opinion in their own words and to relate to specific problems in the reviewed paper. Considerable judgment is required from the reviewer to find the proper balance between making suggestions to improve the paper while respecting the authors' prerogatives.

The free-form evaluation typically contains two sections. The first is an overall evaluation, which comments on the overall suitability of the paper in terms of novelty and originality, significance, correctness, context and presentation. This section should answer the following questions: Does the paper present new results? Are the results important and if so, why? Is the methodology appropriate for the problem addressed? Is the literature review adequate and are the results compared with previous theoretical and experimental results? Is the paper free from errors? Is the presentation adequate in terms of clarity, organization, style, and language? Finally, this section concludes with a bottom line recommendation: to publish as is, to publish with mandatory or suggested revisions, or to reject.

The second section contains specific comments. These comments should be numbered, and keyed to specific locations within the paper, i.e. by page and line numbers, or sections. Every general negative comment in the first section should be backed up with specific comments in this section. Constructive and helpful comments, which suggest how to fix a problem and improve the paper, are far more useful than comments that only point out a problem.

3.3.3 Marked-up manuscript file

In the days when the editorial process manually handled hard copy manuscripts, reviewers would commonly note minor corrections (e.g. mathematical, grammatical, and spelling) directly on the manuscript and short comments in the margins, which were required to be wide to accommodate them. The modern equivalent is to apply these corrections and comments to the electronic file. This can most conveniently be done on the source file (e.g. *.docx) rather than on the pdf files normally supplied to reviewers online. The source files may be requested by the reviewer from the editor. Some editors prefer input on the source document.

To ensure maximum effectiveness, the following guidelines are suggested:
- Before opening the file, change the User ID to something like "reviewer" to protect your anonymity.
- Engage the "track changes" feature so that the suggested revisions will be clear to the editor and to the authors. This will allow the authors to accept or reject each suggested revision.
- Enter marginal comments using the "comments" feature, rather than inserting them directly in the text.

Commenting and correcting directly on the manuscript file has several advantages:
(1) It is efficient and quick for the reviewer, and for the authors, as all of the review can be handled from one file.
(2) The exact location in the text of each comment is accurately displayed.
(3) Suggested corrections to the text are more likely to be accurately transmitted to the final paper.

3.3.4 *Cover letter*

If the review is not transmitted to the editor via a website, the reviewer should ensure that the questionnaire, the free-form evaluation, and the manuscript file with reviewer revisions and comments do not contain any identification of the reviewer (e.g. via user name or author identification in the file information). The reviewer identifies himself to the editor via a cover letter. The cover letter is a short "business letter" whose body can be a single sentence (e.g. "Enclosed please find my review of m/s 12345"). The cover letter should be written on the letterhead of the reviewer's institution, or its electronic equivalent, and include e-mail address and telephone numbers.

3.4 Evolution of Scientific Publication

The form of scientific publication has evolved greatly from hand-written letters addressed to trusted colleagues in Archimedes' time, to the great printed journals of the 18th through 20th centuries, and to the rapidly evolving online web-based scene dominating the early 21st century. Whereas Darwin was paid handsomely to write journal papers in his day, presently authors' institutions are expected to bear an increasing share of publication costs. Most printed journals are also distributed in electronic form, which has the advantage of economically supporting color and multi-media enhancements. Totally electronic journals are developing, and the concept of "open access" is encouraging journals to develop new financial models. Open access is further promoted on the basis that tax-funded research results should be freely available to all tax-payers.

Furthermore, academics are being increasingly evaluated on the basis of various "metrics" — the "impact factor" of the journals in which they publish and of their own papers, their personal "h-factor", etc. In light of this rapid evolution, how and where should a young researcher publish? Our advice is to ignore fads and fashions, whether it is the latest gimmick or a new metric, and concentrate on one consideration: "where are the papers which most influenced your work published?" The researchers who published there most likely read what is published there. This is the community that you want to reach and to influence, and it is this community which will ultimately determine the worth of your work and your standing. Try to publish where these researchers publish and presumably read, and don't worry about the latest fad or metric.

Chapter 4

Conference Presentations: Lectures and Posters

4.1 Definition and Scope

Conferences, symposia and workshops are scientific meetings with formal sessions in which the participants convey new results. They provide informal opportunities for the participants to become personally acquainted with each other and to network, forming contacts for employment and collaboration. In pure research fields, almost all of the participants are researchers and the program is wholly scientific. In more applied fields, participants are likely to include industrial and commercial practitioners as well. Conferences with topics of more general interest, particularly in the social sciences, may also attract members of the press and the general public. Some conferences also include a technical exhibition, in which companies and other organizations have booths which display their products and services. However, scientific conferences are primarily organized to convey scientific results and are distinguished from trade shows, which are primarily for promoting sales.

This chapter gives guidelines for preparing and presenting scientific work through lectures and posters at conferences. The section on lectures is equally applicable to other oral presentations of research results, for example in departmental seminars or company review meetings. This chapter also discusses conference conduct, including what goes on between the formal sessions. It concludes with some guidance on organizing a conference.

4.2 Lectures

4.2.1 The lecture scenario

In a typical conference, the bulk of the time is scheduled for formal lectures. The time allotted per lecture is usually quite short, perhaps 30–60 minutes for one or a few plenary or keynote talks, 20–40 minutes for invited talks, and only 10–20 minutes for most submitted papers. Even in other settings, such as an academic departmental seminar or a company review meeting, the time allotted for a lecture is short (typically an hour slot, of which 40 minutes is allocated to the lecture itself). Furthermore, in a conference, participants have probably been listening to other presentations for hours or even days, in dark stuffy rooms, and their patience, alertness, and general ability to absorb new material may be far from optimal. The novice researcher who wants to present all the details of his multi-year project to the world will definitely be challenged. In fact, it cannot be done. This section will discuss what can be done, and how to do it effectively.

4.2.2 Lecture preparation

Preparing a lecture comprises four inter-related activities: (1) selecting and organizing material, (2) preparing graphics, (3) rehearsal, and (4) fine-tuning. The paragraphs below will discuss each in turn.

4.2.2.1 Selecting and organizing the material

The key to presenting a good lecture is to select material appropriate to the audience and to the time available. The first issue is "who is the audience?" At a departmental seminar, participants may come from a wide variety of specialties. A university mechanical engineering department may have members specializing in mechanical design, strength of materials, robotics, control, power, fluid mechanics, etc. In contrast, the attendees at one of the very specialized parallel sessions in a specialized conference might have much narrower focus. If the attendees include both researchers and practitioners, material should be included to interest both to the extent possible.

Time allotted for the lecture is the critical issue. Any thought of presenting a project in its entirety should be dismissed *a priori* as it is utterly impossible. Instead, select aspects of the project that are apt to interest the audience, in an amount suitable for the time available. The time available is the net time allotted to the talk, after subtracting time for switching speakers, and for questions and answers from the audience. Thus, for example, if 15 minutes per talk is scheduled in the program, probably only 11–12 minutes will be available for the lecture itself.

> **Researchers and Practitioners**
>
> A researcher produces new knowledge. The practitioner uses that knowledge to do something useful. Researchers are often associated with a university, institute, or R&D department of a company. Practitioners include design, manufacturing, sales, and service personnel, as well as their managers. They might come to a scientific meeting to learn what's new, with the aim of incorporating new findings into practical products or services.

The organization of a lecture is similar to that of a research report, i.e. Introduction, Methodology, Results, Discussion and Conclusions. However, the limited time, and the limited ability of the audience to absorb and remember detail, dictate emphasizing the main points.

The Introduction should convey what the overall problem is, keeping in mind the expected composition of the audience, and the objective of the work. There is no need, nor time, for a literature review. One or two very important prior works might be mentioned, but a complete citation is unnecessary — the audience will not remember details such as page or volume numbers. Giving an outline of the talk is only appropriate for longer talks, or talks where the organization is not standard. However, giving the audience a preview of the key point is useful as it alerts them to focus on it during the main presentation. The repetition also improves memory retention. The Introduction should take about 10–20% of the allotted time.

In a short lecture, the Methodology section is abbreviated. Only the most important aspects are presented — what is necessary to understand the results. There is no expectation to hear sufficient detail to duplicate the results.

Most of the time should be devoted to the Results, and the explanation and implications of the results. As in the research paper, the lecture ends with Conclusions, which should include the answer to the research question, and perhaps one or two other points which the author hopes that the audience will remember. Conclusions should repeat points made earlier in the lecture. By telling the audience in the Introduction what they will hear, telling them in the Results and Discussion sections, and finally in the Conclusions summarizing what they were already told, the main points are presented three times, considerably improving the probability that the audience will remember them.

> **Humor?**
> Should you try to lighten the atmosphere and grab the attention of your drowsy audience with a joke? Probably not — at least not with an international audience. Humor is very much culturally bound. A joke which will have one nationality rolling on the floor with laughter is apt to be considered by another as lacking humor, or even as offensive. Jokes which appeal to the culture of science and technology might have a better chance. But it is probably better to use the short time available in a conference lecture to stick to the main points, and save the jokes for private conversations with colleagues sharing similar culture.

4.2.2.2 Graphics

The lecture is almost invariably accompanied by slides projected onto a screen in front of the audience. The slides:
(1) present visual material such as photographs, graphs, and other illustrations,
(2) focus audience attention towards the front of the auditorium,

(3) help newcomers, and those whose attention has wandered, to get synched-in with the lecture and
(4) complement the presenter's speech, and remediate audio transmission problems, e.g. language or other presentation problems of the lecturer, poor amplification or other acoustic problems, and hearing or linguistic problems of the audience.

Guidelines for preparing graphs for slide presentations are explained in the following paragraphs, summarized in Table 4.1 and illustrated in Figures 4.1 and 4.2.

Each slide should be titled and numbered. The title allows the listeners to reconnect if their attention wanders. The number allows listeners to request redisplay of a specific slide for questions and discussion.

As a rule of thumb, approximately one slide per minute is needed. This can vary considerably according to the nature of the slide, and how it is used, from about 0.5 minutes for a photograph requiring little explanation to 2 minutes for a very 'busy' slide requiring considerable explanation.

An application such as "PowerPoint" specifically intended for slide preparation should be used, with its default options for margins, font sizes, and organization. The file format must be compatible with the computer available for screening at the conference. Letter and other feature sizes must be sufficiently large to be legible at the back of the auditorium. The minimum font size, e.g. for a subscript, should be 14 points. Titles should be 40–48 points, and normal text 20–24 points. There should be plenty of space between lines of text. Use the font sizes and spacing suggested by the slide preparation programs as their default option.

Choose a plain white or very light pastel background and dark letters. Colored backgrounds, while aesthetically pleasing, are usually less legible, especially under difficult light conditions. It is not uncommon that the projector is insufficiently strong or the background light too strong. Avoid red lettering as it does not project well.

Table 4.1 Lecture slide guidelines.

Issue	Guideline
Audience focus	Always display a slide Title and number slides
Amount of material	~ 1 slide per minute
Visibility and legibility	Use slide program defaults. General guidelines: • Titles — 40–48 points, • Main text — 20–32 points, minimum (for subscripts) — 14 points • Wide spacing between lines • No workshop drawings • No scans from books or journals without magnification and emphasis

Color	• White or light background
	• Default text — black or other dark color
	• Use color for title and emphasized text
	• In figures, use different colors to differentiate:
	   o subsystems
	   o materials
	   o curves on graphs

Text	Key words only — no complete sentences
Equations	Avoid them If displayed, equation should be titled, all terms defined
Diagrams and photographs	Label key features Explain key features verbally Provide and point-out scale of diagrams and photographs
Animations and clips	Upload and test on conference computer well in advance Have a back-up plan
Last slide	During Q&A, display either conclusions or interesting graphic (not "thank you")

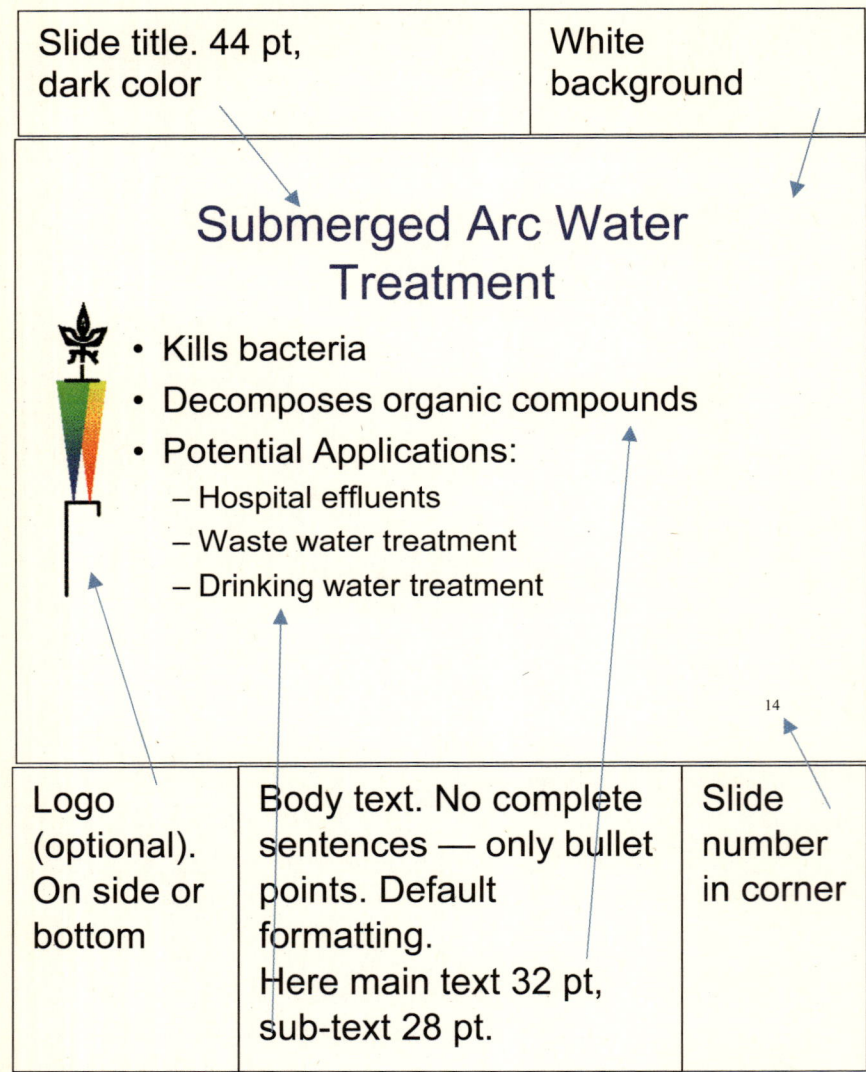

Figure 4.1. Sample text slide.

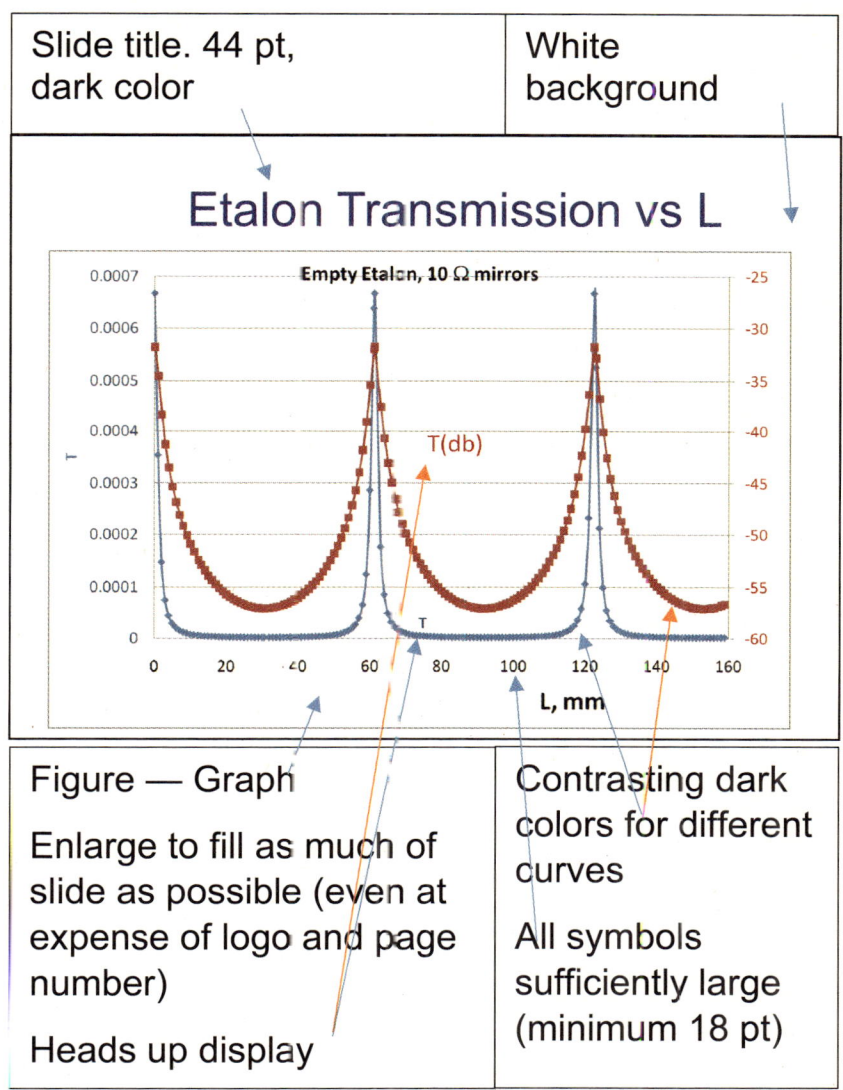

Figure 4.2. Sample graphic slide.

Text on slides should be comprised only of key words. Complete grammatical sentences should never be projected. The purpose of textual slides is to keep the audience focused on the words delivered by the lecturer. We want the audience to pay attention to the lecturer, not to concentrate on reading text from slides.

Even in a theoretical paper, avoid slides containing only equations. If equations are projected, they should be named (e.g. "conservation of momentum") and labeled, and all the symbols defined, both on the slide and orally. In keeping with emphasizing the result and minimizing the method, long detailed derivation of equations should be avoided in lectures; rather the results and their implications should be emphasized.

Prepare graphical material such as photographs, diagrams, and graphs with feature sizes sufficiently large to be viewed comfortably from the back of the auditorium. Line widths and font sizes suitable for print might not be suitable for projection. Workshop drawings are too detailed and the lines too thin to be legible. Likewise, drawings and graphs copied from print will not be legible. Important features should be both labeled on the slide, and pointed out by the lecturer. Photographs and micrographs are best displayed singly, and should fill the slide except for space reserved for titles and labels. Placing multiple photographs on a single slide should be avoided, as the feature size is usually too small and more difficult to discern, especially in the back of the auditorium. Placing multiple photographs on a single slide is only appropriate if the comparison between the photographs is the critical point, and if the key features are sufficiently large to be easily viewed from the back of the auditorium.

During the question and answer period after the lecture, display a slide with interesting content, not a "thank-you" slide. Possibilities include the main conclusions or an interesting graphic.

Animations and film clips can be very effective during a lecture. All of the previous comments concerning feature size and legibility apply to clips and animations. In addition, care must be exercised with embedded material. When the file is transferred to the auditorium computer, and stored in a different directory, it is likely that the embedded files will fail to play properly. Better results are usually obtained by using the "show" file options (i.e. save as *.pps, *.ppsm, *.ppsx files in PowerPoint). Even so, it is wise to upload the file to the auditorium computer well in advance and to verify its proper operation. Have a back-up plan in case the file does not operate properly: to do without the animation, to use your own laptop computer or to go out of the slide program and into the source program for the animation or clip. In the latter two cases, be prepared to cut material to allow for the extra time in switching programs or computers.

4.2.2.3 Rehearsal

Rehearse until you can smoothly deliver the lecture without referring to a script or notes, other than the lecture slides. Reading a lecture is a sure recipe for putting half of the audience to sleep. Most audiences in scientific conferences, even when organized by a local or national organization, tend to be international due to the global mobility of researchers. Accordingly, use simple language, commonly used terms, simple grammar, and avoid acronyms that may not be well known.

Initially the rehearsal can be alone, or in front of a mirror. Recording and then listening to the rehearsal can help you discover and correct delivery problems. But the most helpful final rehearsal is a mock seminar in front of colleagues, who time the presentation and give you feedback both on content and delivery. Thorough rehearsal is even more critical if you will be lecturing in a language other than your mother tongue.

4.2.2.4 Fine-tuning

Almost inevitably, timing the lecture delivery during rehearsal will reveal that the original draft of the lecture is too long. The required action is **not** to speak faster! Under the psychological pressure of a real audience, this strategy rarely works. Furthermore, speaking slowly and clearly is absolutely necessary for the lecture to be understood. The best approach is to shorten the presentation. Review the slides, and eliminate elements of background and methodology that are not essential for understanding the results. If this is insufficient, the next step is to examine and prioritize the results. Eliminate the lowest priority results, until the talk fits within the allocated net time. This process may require several iterations of cutting and re-timing the lecture.

4.2.3 Lecture delivery

The lecture should be delivered while facing the audience, making eye contact with the participants, and speaking to the audience, not reading to them. This delivery creates a psychological bond between the lecturer and the audience, and helps keep audience attention.

Using a laser pointer makes it very difficult to follow this advice. It is preferable for the lecturer to point at relevant features of the slides using an on-screen cursor, controlled by the mouse. Ideally, a large dark cursor should be chosen. If the slides are presented via PowerPoint, before beginning the lecture, right click on the screen, and choose *pointer options / arrow options / visible*, to keep the cursor visible at all times. Alternatively, use the pseudo-laser spot as a cursor, if available (e.g. in PowerPoint 2013 and 2016). It is preferable for the auditorium computer to be mounted on the lectern to allow the lecturer to glance slightly downward towards the computer screen as needed, while generally facing the audience. The advantages of a screen cursor over a laser pointer are summarized in Table 4.2.

Table 4.2. Advantages of pointing with the cursor rather than a laser pointer.

Issue	Laser Pointer	Cursor
Eye contact	Presenter must turn away from the audience, cannot keep eye contact.	Presenter always faces audience — glances down at lectern computer screen when needed.
Steadiness	Laser beam constantly darts around — distracting.	Cursor can be placed accurately and steadily on the desired feature.
Speckle	Laser speckle is annoying and distracting.	No speckle.
Multiple screens in large auditorium	Speaker can only point at one screen at a time with laser pointer. Participants far from that screen are disadvantaged.	Cursor is simultaneously projected on all of the screens.
Availability	Pointer can "walk away".	Mouse or other pointing device is always connected to computer.

A lecture is analogous to a multi-channel, multi-media broadcast communications link (Figure 4.3). Part of the information is conveyed in each channel, and when properly summed at the receiver, fuller information is provided. However, noise in any one of the channels can distract the human receiver and reduce the information received.

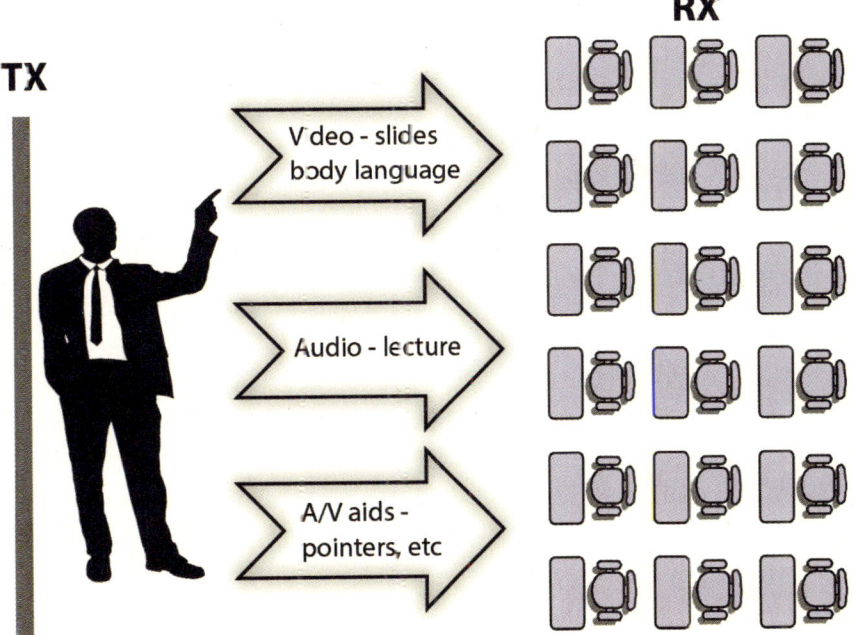

Figure 4.3. The lecture as a multi-media communications channel. The lecturer functions as the transmitter (TX), and the members of the audience function as multiple receivers (RX).

To integrate the "audio" and "video" channels, the lecturer can point to the relevant portion of each slide while speaking. When presenting graphical elements such drawings, graphs, photographs, clips, etc., describe important features orally, and point them out using the cursor and mouse.

Speak slowly and clearly. Controlling voice volume is important. If an audio amplifier is used, the lapel or hand-held microphone must be correctly positioned, and the volume adjusted, in order that the speaker can be clearly heard in all parts of the auditorium. The speaker must control the inflection and volume of his voice. A downward inflection (decreasing audio frequency) orally denotes a period (.), while an upward inflection denotes a question mark (?). Some speakers have a problem maintaining amplitude during the course of a sentence with the result that the ends of sentences are not audible. Some common lecture delivery problems, and their cures, are summarized in Table 4.3.

Table 4.3. Common lecture problems and their cures.

Problem	Cure
Lecture too long.	Choose material according to net allotted time. Time rehearsal, and cut material as needed. ~ 1 slide per minute.
Illegible graphics.	Use sufficiently large font and feature sizes, and space between lines. Prepare slides with a dedicated slide preparation application, and use its default font sizes and spacing. Avoid placing multiple figures on single slide.
Facing away from audience, not maintaining eye contact.	Do not use laser pointer. Look at slides via computer screen on the lectern. Point with mouse/cursor.
Ragged delivery — searching for right word.	Rehearse until delivery is smooth. If unable to deliver smoothly, request poster presentation.
Voice fading or incorrect inflection at end of sentence.	Record/playback rehearsal or rehearse in front of colleagues to identify problem. Rehearse to eliminate the problem. If unable, request poster presentation.
Audio blackouts when not facing microphone	Use correctly positioned lapel mike if available. If using lectern microphone, do not turn back on audience and microphone — don't look at the projection screen. Instead always face the audience and microphone, look at slides only via lectern screen, and point with cursor/mouse.

4.3 Posters

Poster sessions wherein authors display and stand by a poster presenting their work are increasingly popular features at scientific conferences. They are attractive to conference organizers because they can efficiently allow a large number of participants to present their work simultaneously. They are advantageous to participants because they facilitate in-depth discussion between presenters and the typically few participants who are really interested in the particular topic of the poster. The paragraphs below will present some guidelines for preparing and presenting posters.

4.3.1 Poster preparation

Poster preparation includes preparing the poster itself and preparing a two-minute verbal summary of the poster. Posters are typically 120×90 cm or A0 (118.9×84.1 cm) — the exact size allocated is usually stipulated by the conference organizing committee. The most attractive posters are printed full-size either on poster board or cloth, usually using special large ink-jet printers. Poster board is less expensive but must be rolled-up for transport. Cloth is more expensive, but it can be readily folded and transported inside a briefcase or a backpack. The poster can also be assembled from individual sheets of paper, each sheet having a similar composition as a slide for a lecture. This is quite inexpensive, as the sheets can be readily printed on an office or home color printer. However, more time is required to position and hang the individual sheets, and the end result is not aesthetic.

Presenters should **never** simply post a copy of a printed paper, whether enlarged or not, as a poster. No one wants to read a paper standing up, and waste time that could be spent interacting with other participants.

The poster should be designed to serve two audiences: (1) For the casual passers-by, who will merely glance at the poster, possibly when the presenter is not present, an abstract and conclusions in big type convey the point of the research in 10–15 seconds. (2) More importantly, for the few really interested participants, the poster serves as a convenient background for in-depth discussions with the presenter.

The poster itself is best prepared using a presentation program such as Microsoft PowerPoint. The top of the poster should include the poster title, authors, and affiliation, much as it might appear at the beginning of a journal paper, or on the title slide of a lecture. Generally, the organization of the poster is the same as the lecture, and the guidelines for preparation of the lecture slides apply also to the poster, with two exceptions: (1) A bullet-point abstract, in large type, should be presented to serve the needs of the casual passers-by. (2) More detail may be shown, because there will be sufficient time for the presenter to explain those details to the few really interested participants. This may include, for example, more complete and detailed sets of equations, and more details of the experimental apparatus. Nonetheless, choose the material presented on the poster to emphasize answering the Research Question.

As in lecture slides, text should only consist of bulleted key words – no complete sentences. Feature and font sizes should be sufficiently large to be comfortably viewed at a distance of 1 m. To the extent possible, convey your message with graphics, keeping in mind the adage that "a single picture is worth a thousand words". A sample poster is shown in Figure 4.4 below.

Poster presentations are excellent opportunities to show three-dimensional material, and to show video clips, e.g. on a laptop, tablet, or smart phone. If you have such material, ask the conference organizer for a small table upon which to place them.

Prepare a two-minute talk summarizing the main points of the poster. Rehearse and give a practice presentation to a few colleagues, using the poster as a backdrop.

4.3.2 *Poster presentation*

Hang the poster at the time specified by the conference organizer. You should be present at the poster session, and standing adjacent to your poster during all the times scheduled in the conference program. The best way to succeed with a poster presentation is to be proactive. When attendees pass by your poster, greet them and offer to summarize your work (e.g. "May I show you the main points of our research?"). Present your two-minute summary (mentioned in section 4.3.1), pointing to the relevant parts of the poster. At the end, invite questions, and continue to discuss as long as the other person is interested. If, while discussing with some attendees, others come by, invite them into the discussion, by briefly summarizing, in one sentence, the topic currently being discussed.

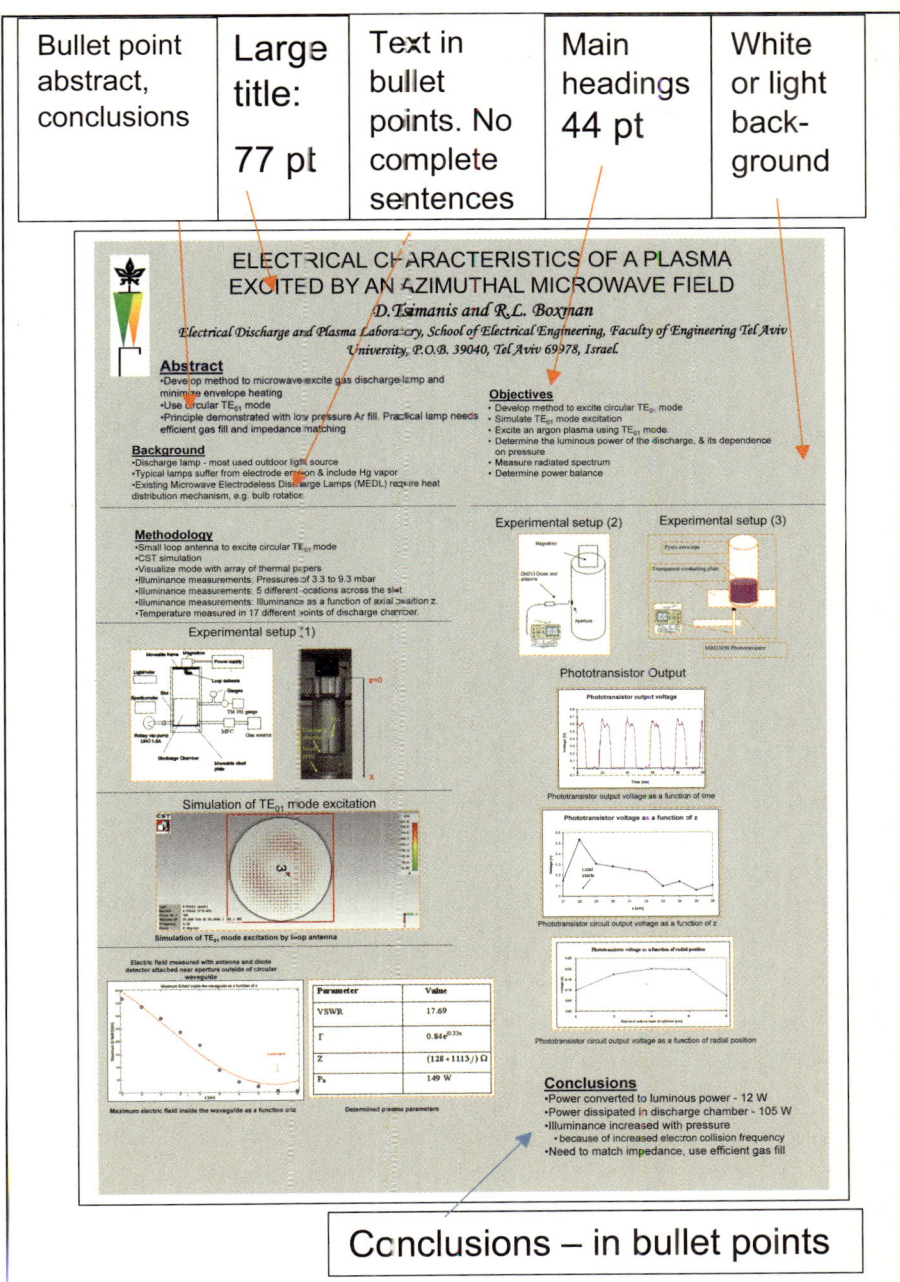

Figure 4.4. Sample poster.

4.4 Conference Conduct

The most important conference activity is personal discussion with colleagues, using the opportunity to learn latest trends, to make contacts for obtaining employment or employing others, and to arrange collaboration (i.e. networking). Accordingly, the most important sessions of the conference are the coffee breaks, meals, and other social activities which facilitate informal interaction. The formal structure of the conference with its lecture sessions provides a framework for these informal interactions. Therefore, you should exploit every opportunity to make new acquaintances at the conference, and renew old ones:

- Ask relevant questions during the discussion period after lectures. Introduce yourself and your affiliation, ask your question, and sit down and listen to the reply. Don't monopolize the discussion period — any follow-up should be during the coffee break.
- While queuing for registration, refreshments, etc., introduce yourself to your neighbors and learn about their work.
- Introduce yourself to participants whom you would like to meet, e.g. authors of papers you have read and enjoyed, potential employers and collaborators, etc. However, don't be a 'pest' — be polite and don't interrupt conversations.
- Arrange in advance to meet colleagues for meals and social activities not organized by the conference (e.g. a drink together, walk around the city, movie, etc.).
- Avoid hanging around with colleagues from your own institution or region, whom you could meet easily at home, and instead use the opportunity to meet new colleagues from distant locations.

It is useful to have business cards (name cards) available to give to new acquaintances. If your institution does not provide you with professionally printed cards, self-designed cards printed using a color printer on card stock are fully adequate for graduate students. (Card stock is available at most stationary stores, manufactured by Avery, Staples, etc.) The card should contain the person's name, title, and all communications data (address, e-mail, phone and fax numbers, and website).

Be aware that in East Asia politeness requires adhering to a certain protocol when exchanging business cards. The card should be presented with two hands, in the direction so that the recipient can read it. Likewise, an offered card should be accepted with two hands, and immediately read. The card should be placed in a card case, or in a jacket or shirt pocket, but never in a rear pants pocket.

Conference attendees, and especially students looking for post-doc opportunities or employment, should always be neatly groomed and attired according to the dress code suggested by the conference organizers. Unless otherwise stipulated, "business attire" is recommended whenever presenting a lecture or poster. For the men, this means a coat and tie, while for the women options are more varied, but a professional look is appropriate.

Organizers invest a lot of time and effort to put together a good conference. Participants should cooperate with their efforts and do everything possible to make their job easier. Follow instructions for submitting abstracts, and make registration and hotel accommodations in a timely manner. If after submitting an abstract to a conference you find that you cannot attend, notify the organizers immediately. Being a no-show, i.e. not appearing for a scheduled presentation without prior notification, can harm your reputation, as well as wreak havoc with the conference program.

4.5 Organizing a Conference

Organizing even a small conference is a major undertaking requiring much effort and involving many aspects including finance and securing sponsorship, venue selection, as well as program organization. Often students are recruited to help in organizing conferences, and it is a task frequently undertaken by researchers early in their career as it gains them significant recognition. This short section will discuss a few aspects of conference organization which are often neglected. When neglected these aspects can significantly and adversely affect either the formal or informal transfer of information.

4.5.1 Scientific team and program

Although all the logistical arrangements are important to a successful scientific conference, the focus on science must be maintained. At the earliest stage, recruit a well-balanced scientific team to organize the program. The team members should be both knowledgeable about the science in the conference field, and about the people making the science. As early as possible, arrange for keynote and invited speakers, including both well-known leaders who draw participants and up-and-coming researchers who are performing breakthrough work. The team members should be proactive in encouraging paper submissions through personal contact.

4.5.2 Early planning and publicity

A successful conference requires a critical mass of participants in order to facilitate the exchange of scientific information as well as meeting financial goals. Early planning, including invitation of key speakers, allows early publicity, and encourages potential participants to attend. Many international participants need a long lead time to arrange funding, obtain required institutional and national permission, and obtain visas, especially for conferences held in the U.S.A. Avoid referring to the conference only by an acronym — the subject field and headline of all publicity material should be sufficiently descriptive that every recipient of the publicity material understands the subject of the conference.

4.5.3 Briefing

Session chairs and speakers need to know in advance when and where to upload talks to the auditorium computer, the time allocated for each talk, and the time allocated for discussion. The session chair should be briefed on what to do if speakers are overtime, and what to do if there is a no-show (in general, to continue with the next present speaker, if there are no parallel sessions, or give a break if there are parallel sessions). Make sure that the chair is familiar with all of the technical equipment (microphones, amplifiers, auditorium computer, projector, room lighting), who will operate each, and what to do or whom to contact if there is a problem with any of the equipment. Poster presenters and the poster session chair should be briefed on the poster size, when to mount and dismount the posters, where mounting supplies are located, how to find the location of each poster and when the poster presenters should be present at their posters. Conference organizers should collect a Speaker Information Form from each speaker, to gather pertinent information needed for the on-site program as well as the introduction of each speaker. These should be distributed to session chairs prior to their sessions.

4.5.4 Auditorium

Lecture halls, auditoriums or comparable space in a hotel or other venue, and their arrangement, equipment, and personnel, should be chosen so that all participants can hear the lecturer and see the screen, throughout the program. Large rooms may require multiple screens or a powerful projector on a large screen. If the room has columns or other obstructions, the visibility of the podium and screen at each location should be checked before arranging seating. Ceiling mounted projectors are preferred, as are wireless lapel microphones.

The podium should be organized so that the lecturer can face the audience at all times. The podium should include a lectern that supports a lap top computer or display screen at chest level, and a mouse for pointing. The computer used for displaying lecture slides should be set up so that a large dark cursor is displayed for pointing. Consider a raised stage in the center of one end of the room.

The operation of all equipment should be tested prior to the opening of the conference. Spares of all essential equipment should be available. A dedicated audio-visual person may be provided by the venue. Conference organizers should meet with this person ahead of time, exchange mobile phone numbers, and set dedicated times throughout the conference to meet up and check on the equipment. If audio-visual support is not provided by the venue, arrange for a suitably competent person to do these jobs.

4.5.5 Poster sessions

Poster sessions are most effective if the following guidelines are followed:
- Encourage poster presentations. Avoid treating them in any way as second class papers. The same selection quality standards should apply to poster and oral papers. (Oral presentations should be chosen on the basis of overall interest in the topic, and the speaking ability of the presenter.)
- Proceedings or special journal issues should not show whether a paper was presented orally or as a poster.
- Provide all food and drink in the poster presentation hall.

4.5.6 Food service

It is important that any food service within the formal conference program, e.g. coffee breaks and lunch, is provided efficiently. Several parallel food and beverage dispersal locations widely distributed in the hall help minimize queuing time. The arrangement at each location should be such that the flow of attendees through each is unidirectional and bottlenecks are avoided (e.g. place the cups first, then the tea bags, then the hot water).

The dietary concerns (whether for health, religion, or conviction) of the participants should be considered. If possible, the ingredients of each selection should be displayed (e.g. on a menu for sit-down meals, or on a card placed with each serving dish for a buffet).

4.5.7 Encouraging informal information exchange

Recognize that the coffee breaks, and other social activities, are in fact the most important conference sessions! Allocate adequate time to them, and instruct lecture session chairs not to allow their sessions to go overtime into coffee breaks or other informal time periods. These periods should be planned with the same care as the formal sessions, to encourage informal exchange of information.

For further reading:

Michael Alley, *The Craft of Scientific Presentations, Critical Steps to Succeed and Critical Errors to Avoid*, Second edition, Springer, 2013.

Claus Ascheron and Angela Kickuth, *Make Your Mark in Science: Creativity, Presenting, Publishing and Patents — A Guide for Young Scientists*, John Wiley & Sons, 2005

Felice Frankel and Angela H. De Pace, *Visual Strategies: A Practical Guide to Graphics for Scientists and Engineers*, Yale University Press, 2012.

Angelika H. Hofmann, *Scientific Writing and Communication: Papers, Proposals and Presentations*, Second edition, Oxford University Press, 2013

Chapter 5

Research Proposals

5.1 Introduction

A research proposal is a document written to convince an institution to supply resources to conduct a specific proposed research project. In the case of an academic researcher, the institution is typically a public or governmental funding agency (e.g. NSF, NIH, various EU programs) and the resource requested is generally money. In some cases, the resource might be the use of a shared facility, e.g. a space telescope, or an accelerator beam line. Industrial researchers commonly write internal research proposals to fund a project of interest to the proposer: the requested resources might include money, the researcher's time, and the use of company facilities. Graduate students may be required to write a thesis proposal. In this case, the requested resource is the use of the student's time (often supported by a fellowship or assistantship) as well as the time and facilities of the student's thesis supervisor.

It should be noted that in many circumstances, particularly in proposals to public funding agencies, competition for limited resources is fierce. To get funded, it is not enough to write a good proposal — the proposal must be among the best of many competing proposals. Furthermore, most proposals are good, so referees and review committees look for weaknesses in proposals in order to eliminate them.

5.2 The Grantor

In order to decide which proposals to approve or fund, the grantor or his representatives (or university graduate school examination committee or

5.2.1 *Is the research topic worthy of study?*

The answer to this question typically depends on the individual criteria of the fund or program. Some funds favor basic science, so that a topic worthy of study is one which will answer important open scientific questions. Others favor applications, and search for projects which will yield a useful product or process. Some funds have a very broad perspective, whereas others are narrowly focused, and finance projects only within their focus. And some funds issue a number of calls for proposals, each one focused on some narrow field or with some specific requirements. The operative conclusion is that it is wise to carefully study the call for proposals and the background of the granting agency in order to formulate the proposal to answer the specific requirements and concerns of the grantor. Where possible, contact the program manager or point-of-contact of the granting agency. In some cases, they may be able to supply additional background information or discuss your research ideas from the perspective of their needs. Most grantors want to fund groundbreaking, novel projects rather than routine "turning the crank" investigations, but they also want projects that will be successful.

5.2.2 *What is the probability of success?*

The grantor looks for signs that the proposers are likely to succeed (or not succeed). This includes examining:
- The Background section, to see if the proposers are well versed in the field,
- The Methodology and Work Plan sections to evaluate if the methods are suitable for attacking the problem and if the proposed work is logical and reasonable, and
- The Resources and CV sections to consider if the proposed team has the knowledge, experience and track record which suggest a high probability of success.

- The Resources section to see whether the team has adequate resources, other than those requested in the proposal, to execute the proposed research program. These may include equipment either in their laboratory or that is available to them.

5.2.3 What resources are required?

How much money, time, or other facilities are required to achieve the goals stated in the proposal? The grantor is usually trying to maximize his "bang for the buck", i.e. the probable result divided by the investment of the grantor's resources. Hence, between two proposals with the same expected scientific or industrial merit, a grantor will usually favor the less expensive alternative.

5.3 Organization of the Research Proposal

Most granting agencies provide a grant application package. The package usually includes forms for supplying basic information and a suggested structure for the body of the proposal, which is typically limited in length (e.g. 10 pages). The structure described below is typical, and may also be used in applying to an agency which does not provide a form.

5.3.1 Abstract

Much like the Abstract of a research report, the Abstract (or summary) of the research proposal summarizes the main ideas. It is useful to write a quick draft of the Abstract at the beginning of the proposal writing process, and then revise and polish it after the body of the proposal is composed.

The Abstract should include one or two sentences on each of the following: the background of the proposal, its objective, the expected significance of achieving the objective, and the methodology and work plan. The monetary sum (or other resources) requested and the time required for executing the proposed project should also be stated.

5.3.2 Subject Area

The Subject Area section is similar to Stage 1 of the Introduction of the research report (Background). It defines and describes the overall subject area in broad, general terms. The objective of this section is to help the administrative staff of a fund to assign the proposal to the appropriate officer or committee for further handling. Therefore, the target audience is the layman or non-specialist. Technical terms and jargon should be avoided. Typically, this section is ¼–½ page in length.

5.3.3 Scientific or Technological Background

The Background section is similar to Stage 2 of the Introduction of the research report (the Literature Review). In the case of a scientific proposal, it is very similar, whereas in a more technically oriented project, it may resemble a market survey, and cite trade journal articles and patents as well as scientific papers. In either case, the objective is to describe the state of the art at the present time. This section should be used both to convince the proposal evaluators that the proposers are well versed in the subject and to set the background for showing the novelty of the proposed project. All of the guidelines set forth in section 2.4.2 apply here as well.

The logical place to describe preliminary unpublished results of the proposers on the proposed topic is in this section, after all published work (including that of the proposers) is reviewed. Unpublished preliminary results must be clearly identified as such, either by using an appropriate sub-heading, or in the lead sentence of the first paragraph. Some granting agencies, however, specify that preliminary results be presented in a separate section later in the proposal. One solution is to present preliminary results in the logical location, i.e. near the end of the Background section, and to refer the readers to this subsection at the location designated by the agency.

The Background section should end with a paragraph that first summarizes all of the background in one or two sentences, and then presents the gap. The gap is a statement of open questions, or issues which have not yet been investigated. It is similar to Stage 3 of the Introduction to the research report. All of the guidelines in section 2.4.3 apply here, but often there are multiple gaps, and each may require a separate sentence. In the research proposal, it is doubly important that the gap sentences be negative, focused, and specific, and not be "wishy-washy". No evaluator will forgive a proposer who is not well versed in the field (and who writes in a gap sentence ...*to the best of my knowledge*) or who ignores the previous papers closest to what the proposers are proposing *(...few researchers have investigated....).*

5.3.4 Objective

The Objective section states the purpose of the research, and is similar to the Statement of Purpose in Stage 4 of the Introduction of the research report. For a multi-year project, an overall objective is stated first. Then several intermediate objectives are given. The objective is stated in the present tense in the research proposal (e.g. *The objective of the proposed research is to* ...). Most of the guidelines in section 2.4.4 apply here. State the objective using decisive words such as determine, develop, construct, understand, etc. Do **not** write that the objective of the research is to do research (or study, investigate, examine, etc.).

Take care not to mix the objective with the means of obtaining the objective (e.g. *The objective of the proposed project is to stabilize the dynamics of a rocket using a phase locked loop. "Stabilizing the dynamics"* is the objective, *"using a phase locked loop"* is the means.) The means should be presented later, in the Methodology section. If, in fact, the means is the novel aspect, i.e. the objective has already been achieved by some other means, then the objective should be stated as improving the means, developing new means, determining the properties of the means or testing its suitability for a specific application etc. For example: *The objective of the proposed research is to improve the stabilization of rocket dynamics* or *The objective of the proposed research is to determine if using a phase locked loop can improve the stabilization of rocket dynamics.*

> **The Elephants are charging! What's my objective?
> How do I achieve it?**
>
> Confused between the objective and the means for achieving it? Consider the following scenario. You are alone on the African savanna, and suddenly a herd of elephants appear, all of them charging at you! At that moment, clearly your objective is **to save your life**. How? You want to use the most effective available means. You would probably be considered either insane or unreasonable if you said that your objective was "to save your life with a ball point pen."
>
> Likewise, in a proposal, you should state your ultimate objective, in the Objective section, not how the objective will be secured. State how you will achieve the objective in the Methodology section. This should be the best means available to you.

5.3.5 *Expected Results and their Significance*

The Expected Results and their Significance section has similarities to Stages 5 and 6, preview and value statements, in the Introduction of the research report. Expected results can be stated explicitly, if there is a well-grounded prediction based on calculations or some preliminary experimental results. Otherwise, the expected results may be described in more general terms. Usually, the expected results should be presented in one short paragraph.

A second paragraph should explain the significance of the expected results. If the expected results are obtained or the objective achieved, "So what?" Why is it important to obtain this result or achieve this goal? What will you, other researchers, or practitioners be able to do, that they can't do now? Answering these questions specifically, explicitly and directly is very important for getting a proposed project funded. General statements such as *...obtaining these results will be a great advance in the field of xxxx* are worthless. Statements which are specific to the proposed research are required. If you can't explain the significance of your proposed research to the evaluators, most likely your project will not be funded. More importantly, you should be asking yourself, why of all of the myriad possible topics that you could investigate, do you want to investigate this one? If there is no answer, you should really consider a different topic.

5.3.6 Methodology

The Methodology section, together with the Work Plan section that follows, is the heart of the proposal. Here you explain what you propose to do in detail. Many funding agencies group these two sections together into a single Methodology and Work Plan section. However, the text for these two titles have different function and form, so even if included in a single section, the proposer should first deal with the Methodology and then the Work Plan. The Methodology section explains what methods will be used during the proposed research. The Work Plan describes when each one of these methods will be used, i.e. the sequence of actions.

The Methodology section of the research proposal is similar in many respects to the Methodology section in the research report. Hence, most of the guidelines given in Section 2.5 are applicable here. There are, however, two main differences:

(1) In the research report, you report what you did, and therefore use the past tense, whereas in the research proposal, you describe what you will do and therefore use the future tense. Generally, avoid statements in the form of *X can be determined* using *Y method.* This statement only gives general information, but does not unambiguously say that you will use method Y. The "can be" formulation may be used to indicate that there are several choices which the proposer considered, e.g. *X can be determined by methods A, B and C...* But such a statement should be followed by an unambiguous statement, in the future tense, that clearly states your intentions, e.g. *In the proposed research, method B will be used because........*

(2) In the research report, the methodology must be reported in sufficient detail to allow duplication of the results. This is not required in a proposal. Rather, sufficient detail should be given so

that the evaluator understands what you plan to do, and is convinced that you know what you are doing.

As in the research report, methods which are well known may be merely mentioned and supported by a citation to a reference where the method is explained in detail. The proposed choice of conditions might be mentioned, if important. However, methods which are not standard, and especially new or novel methods, should be described in sufficient detail that the evaluator understands your intentions.

5.3.7 Work Plan

The Work Plan describes in what sequence each of the various methods will be employed. The textual description is usually supported by a timetable or a Gantt chart (a bar graph with horizontal lines indicating the span of time in which each activity is conducted). The chart shows the breakdown of the project into phases or tasks over time. The text describes each phase or task and indicates for each which methods (already described in the Methodology section) will be used.

Some researchers, especially beginners, have difficulty formulating a work plan, because each step after the first one might depend on the results of the preceeding step and these results are not known *a priori*. Proposal evaluators are experienced researchers and are well aware of the vagaries of research. Nonetheless, they expect to see a work plan that shows that the proposers have thought things through. The various possible results should be explained, and the subsequent steps which will be taken in each event should be described. If there are multiple decision points, a flow chart showing what will happen in each eventuality will clarify the textual description. In addition to convincing the evaluators of the soundness of your proposal, a well thought-out plan helps keep a funded project on track. Work plans are written with bits and bytes in some computer memory, and not chiseled in stone. During the project, work plans may be modified. Most granting agencies expect changes and have policies for reporting them. However, the original work plan, and subsequent modifications, should be carefully thought out.

5.3.8 *Resources*

The objective of the Resources section is to show the evaluators that the proposers have the human and material resources to successfully execute the proposed project if funding is granted. This section should begin with organizational and human resources. First, describe the proposers' institution in one or two sentences. This may be omitted if the institution is well known to the granter. Then describe the proposers' group in a single sentence, stating its most important mission or activity. Follow with a description of the key personnel. The most important and relevant aspects of the principal investigators' background and accomplishments should be described in 2–3 sentences each. Other key personnel should be described in 1–2 sentences. In each case, mention their role in the proposed project.

In a separate paragraph, list the available facilities relevant to the proposed project. Begin with the equipment available in the proposers' own laboratories, and then continue with facilities available elsewhere in the proposers' institutions, and finally describe any facilities that will be used outside of the proposers' institutions. In this section, you should show the evaluators that there is a suitable person, and suitable facilities, to accomplish each task described in the Work Plan, using the methods noted in the Methodology section.

The above proposal sections, from the Subject Area through the Resources, are typically strictly limited in length (e.g. to 10 pages). It is important that this limited space be used optimally. The Objectives, Significance, Methodology, and Work Plan sections should occupy the majority of the allowed space.

5.3.9 *References*

References cited in the proposal, typically in the Subject Area, Background, and Methodology sections, should be listed here. Although the granting agency does not usually specify the format, the formatting must be uniform within each category (e.g. book, journal paper, etc.). It is good practice to use the format of a leading journal in the field of the proposal.

5.3.10 *Budget*

Proposals requesting a grant of money will generally require a budget which details its proposed use. Usually, the granter provides a form on which to present the budget. The budget example in Table 5.1 below is typical and can serve as a template in cases where the granting agency does not provide a budget form. The budget is organized into broad categories as described below. The specific outlays in each category are presented on separate lines. The expected outlay in each year of the proposed project is presented in separate columns. It is good practice to prepare the budget first on a spreadsheet and then to transcribe it onto the granter's form at the final stage. This allows editing of and adjusting the budget during the proposal writing process.

Each granting agency has its own rules concerning the size of the budget, and what they will or will not fund. Some agencies will not fund equipment, others will fund only specialized equipment, yet others are more liberal. Some agencies will fund the salary of all researchers but some will not fund researchers who are on the permanent staff of the proposing institution. Some agencies expect travel, e.g. between cooperating partners in a multi-national research consortium, others do not fund travel. The agency's budgetary rules should be studied carefully before formulating the proposal, to make sure that the proposal is suitable to the support offered by the granting agency, and certainly before formulating the budget, to ensure that all requested line items are allowed.

Typically, the budget is composed of the following categories:

(1) Salaries and benefits. Usually all participants in the proposed project are listed here, even if the project is not funding them (in which case 0 is listed as the amount). Key personnel should be listed both by name and their role in the project. Less senior participants, who have not been chosen yet, may be listed only by their role in the project. Non-salary support of personnel (e.g. via scholarship or fellowship), if requested from the granting agency, is also listed here. Some granting agencies require detailed breakdowns of man-months, the fraction of time to be devoted to the proposed project, etc.

(2) Equipment. The definition of equipment varies among granting agencies, but generally includes equipment whose cost exceeds some threshold value and which is not consumed or destroyed by use.

Table 5.1. Sample budget and budget justification for a research proposal. The table shows a typical working spread-sheet for working out the budget, with typical categories. The sums used, however, should not be considered representative.

Sample Budget				all figures in US$		
			Yr 1	Yr 2	Yr 3	total
Salary and Benefits	role in project	% time in project				
Prof. A.M. Oxford	principal investigator	40	$50,000	$50,000	$50,000	**$150,000**
Prof. C.D. Guildford	senior investigator	40	$45,000	$45,000	$45,000	**$135,000**
F.N. Coppersmith	senior technician	25	$20,000	$20,000	$20,000	**$60,000**
	graduate student	100	$35,000	$35,000	$35,000	**$105,000**
		total salary and benefits	$150,000	$150,000	$150,000	**$450,000**
Equipment						
Vacuum system			$12,000			**$12,000**
Manometer gauges and control			$8,000			**$8,000**
Mass flow controller			$6,000			**$6,000**
		total equipment	$26,000	$0	$0	**$26,000**
Materials						
Metals, ceramics, polymers			$3,500	$4,000	$2,000	**$9,500**
Electronic components			$2,000	$1,000	$500	**$3,500**
Mechanical components			$5,000	$4,000	$500	**$9,500**
		total materials	$10,500	$9,000	$3,000	**$22,500**
Services						
Workshop			$5,000	$1,000	$1,000	**$7,000**
Physical Diagnostics			$7,000	$18,000	$16,000	**$41,000**
Mechanical Diagnostics			$4,000	$7,900	$6,000	**$17,900**
		total services	$16,000	$26,900	$23,000	**$65,900**
		subtotal	$202,500	$185,900	$176,000	$564,400
Overhead		25%	$50,625	$46,475	$44,000	$141,100
		TOTAL	$253,125	$232,375	$220,000	**$705,500**

> **Budget Justification**
> Budget support is requested for four participants. Prof. Oxford, the P.I., will be in overall charge of the project, and will supervise the theoretical effort. Prof. Guildford, the senior investigator, will supervise the experimental part of the project. Senior Technician F.N. Coppersmith will be responsible for building auxiliary apparatus and maintaining the equipment. A graduate student, to be appointed, will conduct the experiments and develop the theoretical model as part of his thesis research. The salaries of the named personnel are per standard University of Strantinford rates. The sum cited for the graduate student is based on a 100% fellowship.
> The experiments will be conducted in a process chamber to be built on site, using the requested allocations for materials and workshop time. It will be evacuated by the vacuum system to be purchased during Year 1. The manometer and gas flow controller will be used to regulate gas flow in the process chamber.
> Physical diagnostics include electron microscopy, x-ray diffraction, and x-ray photoelectric spectroscopy and will be conducted in the Falconer Material Research Center. Mechanical diagnostics include hardness and brittleness testing, and will be conducted in the McKenzie Mechanics Laboratory.

(3) Materials and supplies. This category includes all consumables, e.g. chemicals, nuts and bolts, biological material, etc. Depending on the granting agency, it may also include small instruments and tools, which are not consumed, but are below the agency's price threshold for equipment, e.g. multi-meters, personal computers, etc.

(4) Services. These might include use of special laboratory facilities, machine shops, communications, etc. In some grant forms, services are included under materials and supplies.

(5) Travel. Tickets, accommodations, and living expenses.

(6) Other. Direct expenses which do not fit into the above categories.

(7) Overhead. This is a payment to the institution for the use of lab space, utilities (electricity, water, etc.), cleaning, maintenance, administration, etc. It is usually calculated as a percentage of the sum total of expenses, or it might be calculated as a percentage of salaries. Some granting agencies have policies setting a limit to the overhead rate which they will pay that may conflict with the institutional policy.

Some experienced researchers suggest preparing a minimal budget, believing that a small budget is more likely to be funded than a large budget. Others suggest inflating the budget, since the funding agency is apt to allocate less than what is requested. Our advice is to request what you estimate to be the appropriate budget needed to achieve the proposed objectives, provided it is within the limits set by the funding agency. Set the research objectives and work plan so that the program can be executed within the allowed budget limit. Reviewers are generally experienced researchers who can detect stingy or inflated budgets; if detected, they will lower their evaluation of the proposer as a researcher who knows how to plan research.

Most research institutions have a Research Authority, Sponsored Research Office, or a similar department which encourages and possibly aids in submitting research proposals, and administrating the budget, if granted. This department usually provides guidelines on salaries for temporary research hires and on overhead, and typically checks the budget for compliance with the guidelines of the institution and the granting agency.

Following the presentation of the budget table, large expenditures and unusual expenditures should be explained in a paragraph titled Budget Justification. Explain: the role of each member of the team, how each piece of equipment will be used, the purpose of each trip, any unusual or large expenditures, and how the figures were calculated or estimated.

5.3.11 CV's of key personnel

Most granting agencies require background information on key members of the project team, but who is included in that team and the amount of detail required vary. Some require a full CV, but most provide a one-page template for providing specific highlights. Some want the full list of publications, while others want some specified number of publications that are most relevant to the proposed project. Some include only the principal investigators while others are interested in all of the key personnel. In short, the granting agency's guidelines should be followed.

5.3.12 Letters of cooperation

Granting agencies considering proposals involving several institutions typically require a letter from each institution. This letter should describe the specific role of that institution in the project, and explicitly state that it agrees to be part of the project as described in the proposal.

5.4 Evaluation of Research Proposals

How a research proposal is evaluated depends on the nature of the call for proposals. The situation with the highest probability of success is what might be called an invited proposal. In this case the granting agency approaches a researcher and requests a proposal on a given topic, and might also suggest an approximate budget and time frame. This usually occurs when the granting agency decides that it needs the research to fulfill some aspect of its mission or work program. In this case, the contact person from the agency will generally champion the proposal through the agency's approval process.

More usual are public calls for proposals. These proposals are highly competitive. Many researchers are competing for limited funds. Public funding agencies will typically assign each proposal to a case officer or committee according to the topic. It is important that the Subject Area section be well written and understandable by a non-expert, so that the proposal will be assigned to the officer or committee closest and probably most sympathetic to the topic. Typically, the officer or committee will request reviews from multiple referees. In very competitive calls, probably only proposals which receive unanimously excellent ratings will be funded. Among generally excellent proposals, it is somewhat random which ones will be selected. Accordingly, it is worthwhile, where permitted, to submit proposals to multiple agencies in the hope that the specific configuration of referees at one of the agencies will be unanimously favorable to the proposal.

In a very competitive environment with many good proposals and limited funding, referees and evaluation committees often look for defects in the proposal. Some common problems are examined in Table 5.2.

Table 5.2. Common problems and cures in research proposals.

PROBLEM	EXAMPLE	ANALYSIS	SUGGESTION
Subject Area text too technical or uses too much jargon.	*The translocation of the 6-kDa subunit (MtATP6), a relatively nucleus-encoded subunit of the F0 part of the F1F0-ATP synthase complex, from the nucleus to the Mitochondria under salt induced stress is of great interest in determining nuclear-mitochondrial communication dynamics.*	Opening sentence uses very technical words which a non-expert is not likely to understand. Write this section for non-experts, so they can determine what the general field is. The suggested text avoids much of the jargon, and defines the technical terms which are used.	*Cells in living organisms contain substructures called organelles. The proposed research investigates how one type of organelle, mitochondria, communicates with the nucleus under salt induced stress.*
Absent, unclear, or "wishy-washy" gap statement.	*Few researchers have investigated the influence of magnetic fields on cathode spot dynamics.* *The influence of magnetic fields on cathode spot dynamics is not fully understood.* *To the best of our knowledge, the influence of magnetic fields on cathode spot dynamics has not been investigated.*	The first two examples imply that some researchers have investigated the topic, or that there is partial understanding. The literature review should focus on what was done, and the gap should pinpoint what has not been done. The last example indicates that the proposers are unsure of their knowledge of the field. What evaluation can they expect from the referee? Solution: write a definitive, focused, gap statement.	*The influence of oblique magnetic fields on ion trapping in the electron relaxation region of the cathode spot has not been determined.* Note: the suggestion narrows the gap hinted in the first example.

PROBLEM	EXAMPLE	ANALYSIS	SUGGESTION
Stating that the objective of the proposed research is to do research, or its equivalent (study, investigate, examine, etc.).	*The objective of the proposed research is to investigate nitrogen uric acid decomposition reactions.*	Stating that the objective of the research is to do research indicates poor scientific thinking by the proposers. It is requesting that the granter provide money for the proposers to play in their lab. Use deterministic words when stating the objective that suggest a definitive end point, e.g. measure, determine, understand, etc.	*The objective of the proposed research is to determine the level of uric acid decomposition in diabetes patients.*
Significance unexplained or poorly explained	*Rapid artificial synthesis of Spinosad will significantly advance natural insecticide technology.*	Avoid generalities. State specifically and precisely how achieving the research objectives will allow addressing further scientific issues, or can be utilized in a product or service.	*Spinosad is a natural insecticide derived from the bacterial species Saccharopolyspora spinosa. Its artificial and rapid synthesis overcomes the most significant stumbling block to industrial production of safe and natural insecticides.*
Inadequate methodology or work program		Sufficient details must be given in the methodology and work program to convince the referees that each of the objectives can be achieved.	

PROBLEM	EXAMPLE	ANALYSIS	SUGGESTION
Imbalanced proposal: Too much space in Background; not enough in Methodology and Work Plan		The Methodology and Work Plan sections show what you will do, i.e. they detail what you propose.	In 10 page proposal, no more than 3 pages of background, and at least 5 pages of Objectives, Significance, Methodology, and Work Plan.
Ambiguous statements in the work plan	*The air velocity can be measured by hot-wire thermometry.*	Not clear if this statement is merely providing general information, or if this is the method which is proposed. Use the future tense to clearly indicate the proposer's intention.	*The air velocity will be measured by hot-wire thermometry.*
In theoretical proposals: lengthy mathematical derivation in the methodology section; inadequate work program		Mathematical derivation shows work already done. A proposal is about work which will be done in the future, if the project is approved.	Avoid mathematical derivations in the Methodology section. Instead, present in the Methodology section what problem will be solved, mention relevant equations by name, mention simplifying assumptions or how the equations will be simplified in the future. Present in the work program the sequence of sub-problems which will be attacked, and mention the methods relevant for each.

PROBLEM	EXAMPLE	ANALYSIS	SUGGESTION
Inconsistency	No method or plan given for some objective. Method or task not directed towards an objective. Requested equipment or material not used in the methodology and work plan. A team member does not have a function mentioned in the work plan.	Referees look for inconsistencies as a way of eliminating proposals from serious consideration. Inconsistencies suggest poor thinking or organization of the proposers.	Build the proposal around the objectives. A method must be given to accomplish each objective. All methods and tasks should be specifically aimed to accomplish a specific objective. All requested resources (e.g. personnel, equipment, material, lab tests) must be directed to specific tasks, and explained in the Budget Justification.

For further reading:

Joshua Schimel, *Writing Science: How to Write Papers That Get Cited and Proposals That Get Funded*, Oxford University Press, 2012

Chapter 6

Business Plans

6.1 Introduction

Engineers and scientists sometimes conclude that a result of their research, or some idea that they have, has commercial potential, and that they themselves should play the leading role in its commercialization. While most such ventures ultimately fail, some of the few successful start-ups are legendary. In some countries, start-ups are a large factor driving economic growth. The desire to see the results of one's intellectual efforts achieve practical implementation, as well as the lure of financial success, are powerful incentives for researchers to plunge into the world of business.

The business plan details a proposal to establish a business activity. It is similar to the research proposal discussed in Chapter 5 in that its primary objective is to convince some body to provide funding. It is different in that the potential funder for a start-up is most often an investor, whereas the potential funder for a research proposal is typically a government agency or other non-profit entity. The secondary objective is to help the potential entrepreneurs evaluate whether the proposed business activity is viable, and if so, to ensure that all the key issues have been addressed. A draft plan summarizing the business concept may serve as a basis for consultation with potential advisors and customers, as well as potential investors. The potential entrepreneur may reach the conclusion that the business activity is not viable through the preliminary work of identifying the crucial issues for business success. However, if the evaluation of the business concept is positive, the business plan provides an initial roadmap for launching a new business on a path towards success.

In preparing a business plan, many first-time entrepreneurs will work with a financial consultant or accountant. Ideally, a partner with business experience will be part of the start-up team. Most likely this consultant or partner will prepare the business plan. This chapter introduces the underlying concepts of the business plan. Its aim is to help the researcher understand the types of questions that need to be addressed and to facilitate dialogue with those involved on the business side.

6.2 The Reader

The business plan is intended to be read by those who will make the funding decision, and their representatives and consultants. These may include private investors, and officials in various sorts of venture capital funds, institutional investment houses, banks, and government agencies. Few, if any, of the readers are apt to have technical expertise in the researcher's field, although some may be technologically savvy, and some may engage an expert consultant to examine the technological portions of the business plan. Generally, they are interested in making money, not in advancing science. And as a rule, they are busy people.

The potential funder will ask questions quite similar to those which agencies funding proposed research ask, i.e.: "How much will it cost?" and, "What is the probability of success?" However, rather than the key question in funding research proposals "Is the subject worthy of study?", the parallel question for investors of various kinds is "What will be the return on investment?" — i.e. "How much will the investor profit?" This is **the** most important question for most investors.

The nature of the technological innovation may be important to the investor. Tech-savvy investors look for cutting edge technologies with market changing potential — the next big product or service. Large corporations may be looking for a solution in the development of a next generation product or for sourcing critical components. It should be noted, however, that large businesses are usually not comfortable in working with unknown start-ups. They are more likely to fund an in-house start-up or related business with a track record.

Some government agencies which fund small business projects, and a few private investors who have a social agenda, may ask additional questions. Government agencies are often interested in how the proposed business activity will affect the local or national economy: Will it create jobs, provide a market for other local companies, provide essential products or services to local companies or residents, or perhaps decrease a country's dependence on imports or provide an export product? A private investor might ask if the proposed activity is compatible with his or her specific social agenda (e.g. to support women or some minority, be environmentally friendly, strengthen national defense, etc.).

6.3 Business Plan Structure

The business plan is organized to answer the potential investors' questions. Typical sections are described below.

6.3.1 *Executive Summary*

The Executive Summary is parallel to the Abstract in the research proposal, and summarizes the key elements of the plan. It is the section which will be most commonly read. Often, busy investors will decide based on reading it whether to continue further. The Executive Summary should be short, e.g. two pages or less, and briefly summarize:
- proposed product or service, and its unique contribution to the market
- potential customers
- competitive edge (and how it will be maintained), including the status of patents and other intellectual property
- in the medical field, regulatory status and plans to obtain requisite approvals
- owners and key personnel (and how they will be retained)
- resources: available and needed
- projected profit (when and how much)

6.3.2 Company Description

The Company Description section briefly describes the entity which will implement the business plan, if funded. It is parallel to that part of the Resources section of the research proposal which describes the research group and its key personnel. However, whereas a research group in a well-established research institution is apt to be known to granting agencies, and have easily obtainable public records in the form of web pages and publications, a new start-up venture will probably be unknown to potential investors. Hence this section must start from zero, and be carefully structured.

6.3.2.1 Overall Company Description

The Overall Company Description will typically open with the name of the company, and a brief statement about its status, line of business, and customer base. In the case of a start-up venture, the company might not have been formally organized yet. If so, this should be stated clearly, and the rest will be expressing intentions rather than established facts. If the company has been established, its legal form of ownership (e.g. privately owned by a single individual, partnership, corporation, etc.) and the principal owners are indicated. The existence of a shareholders' agreement setting out the ownership structure, contributions of the founders and division of intellectual property amongst them may be noted. If the company has a mission statement, it may be presented here. A mission statement is one or few sentences, typically less than 30 words, explaining the company's reason for being and its guiding principles.

6.3.2.2 Key Personnel

The Key Personnel section begins with a brief (two to four sentence) description of the background of key personnel and their intended role in the activities proposed by the business plan. Provisions of a shareholders' agreement that demonstrate the commitment of the founders to the ongoing operation of the company may be presented. More extensive résumés of the key personnel are generally appended to the business plan.

6.3.2.3 Current Business Activities

If the company is already established and has existing business activities, these should be briefly described in a Current Business Activities section. Include the nature of the activities (development, production, sales, etc.), its product or service, customers, and revenues.

6.3.2.4 Core Competencies

The Core Competencies section describes briefly the company's strengths. For an established company, this relates to existing activities. For a new company, this section states what the start-up team should be able to do given the background of the team, and any experience they have already in working together.

6.3.3 Products and Services

The Products and Services section describes the products or services that the proposed business activity will offer. Unlike the value statement in the research paper or proposal, innovations and technological advances should be explained in layman's terms. The competitive advantage of the proposed product or service is crucial. Why would potential customers prefer the proposed product or service to other available and potential alternatives? The answer will probably include a description of unique properties or features, ability to maintain higher quality, and/or the ability to profitably underprice the competition.

This section should describe how this advantage will be maintained over the long haul. Critical issues are protection of intellectual property (IP), and the strategy to avoid technological obsolescence. IP includes trademarks, trade secrets, and patents. The status of any IP, as well as the potential to create additional IP, should be noted. In particular, patents and pending patents should be noted, as they often are viewed very positively by investors. (Patents will be discussed in Chapter 7). IP strategy should be outlined here, including the following issues: (1) To what extent will new IP be kept as a trade secret, or disclosed in patent applications? (2) In which countries will patent applications be filed?

In the medical field, obtaining the requisite regulatory approvals is a crucial issue. Indicate the regulatory path the product will need to pass, the status of any regulatory filings, and the plan for obtaining requisite approvals.

6.3.4 *The Market*

The ultimate success of a business activity is measured by the ability of the company to profitably sell its product or service. This section of the business plan must first of all demonstrate that the entrepreneurs understand the potential market, and second that the proposed product or service will fulfill an unmet need in the market. It serves a similar function to the Literature Survey and the Gap statement in the Introduction of the research paper or the Scientific Background and Gap in the research proposal.

The market analysis typically starts with an overall description of the market, focusing on the final customer. It will answer the following questions:
- Who will buy or use the proposed product or service?
- What are the links in the supply and distribution chain?
- Who makes the decision about which products reach the market?

The potential investor will be interested in the size of the market, both in terms of units sold and overall revenue generated. At what price can the product or service be sold? A key question is what fraction of the market can be captured by the proposed product or service. Likewise, the investor will be interested in market trends: growth (or decline) of the overall market, changes in customer preferences, and recent technological and marketing innovations. Sometimes, the technology is so innovative that market potential is the main issue.

A new product or service will need to penetrate an existing market or create a new market in order to succeed. Part of the market description is a survey of the currently available products and services, and the organizations supplying them. Barriers to market penetration should be explained. These may include high capital cost, control of the market by a few very strong players, and regulatory considerations. The latter can be a major issue for medical products.

Knowing the customers and their needs is critical to the success of every business. Part of this knowledge may be garnered from formal market surveys, contracted to specialists, or conducted through literature and web sources. However, direct contact with the potential customers is important. Academic researchers are often remote from the customers, and may need to make a focused effort to reach them, to present the product concept, and to obtain their feedback. Initial contacts can be made at equipment exhibitions that are often attached to scientific conferences and at trade shows. But ultimately the entrepreneurs should visit customers at their premises. For many technological products or services, this will be at other companies who are potential direct or intermediate clients. However, for consumer products, consumers should be addressed directly. The integration of a formal market survey with the anecdotal information garnered by direct contact with potential customers provides the basis for preparing this section of the business plan, as well as forming the best strategy for success.

A description of how the proposed product or service will fit into the market concludes this section. Will it be competing head-to-head with the key products of the main players? Or does it have a realistic chance of meeting an unfilled need in some potentially lucrative niche? A table that highlights the advantages of the proposed product in comparison to competing products can useful in summarizing the market situation and in showing which niche the proposed product would fill.

6.3.5 *Marketing and Production Strategy*

Following the background provided in the preceding sections (i.e. the company's strengths, its technology, products, and services, and the market), the Marketing and Production Strategy section presents the proposed business concept: what products or services will be offered, how will they be produced, and how and to whom will they be sold? Part of the strategy is how to overcome any market entrance barriers noted in the Market section. For medical products in particular, the strategy for obtaining required regulatory approvals must be included here. This section serves much the same function as the Methodology section in the research proposal.

Typically, an industry has a multi-layered structure (known as the production chain) which includes R&D, production, several layers of distribution and sales (wholesale, retail) and customer service. Given the background laid out in the preceding section, how will the company get its product to the ultimate customer? In which links of the production chain will the company participate directly, and in which will it rely on strategic partners? Will the company develop a new product, and sell or license the know-how to an existing manufacturer? Or, more rarely, will the company manufacture the new product itself (and if so, in its own facilities, or by outsourcing, or by some combination), and sell its production to a high-level participant in the distribution chain? Or will it market directly to the ultimate customer and if so, how? (e.g. with a large conventional sales force, or a smaller sales force only via the web?)

6.3.6 Operational Plan

The Marketing and Production section presents the strategy. The Operational Plan explains in detail how this strategy will be carried out. This section has many similarities to the Work Plan section in research proposals. The Operational Plan details the personnel and facilities which will be used to conduct each activity mentioned in the Marketing and Production Strategy section. It outlines how they will operate. This should include assigning resources for executing the strategy for overcoming market entry barriers. What steps will be taken to obtain regulatory approval for medical products? How will the start-up protect its IP? For an early stage start-up, the Operational Plan will necessarily be sketchy. Nonetheless, its existence indicates that the entrepreneurs have thought things through, and are at least aware of what lies ahead. Companies seeking funding to get beyond this initial stage will need to present a more detailed Operational Plan.

6.3.7 Management and Organization

The Management and Organization section describes who will control all of the operational activities described in the preceding section and how. For a small start-up company seeking initial funding, and perhaps only engaging in R&D, this section can be quite brief, but should state who is responsible for what, and who will make critical decisions. However, for a company which will have more than 10 employees, an organization chart should be included, indicating who reports to whom.

6.3.8 Financial Plan

The Financial Plan ultimately presents the "bottom line" of the business plan — what is the amount of investment required, how much money the investors will see as profit, and in what form (e.g. increase in the value of the equity stock which they hold, dividends, etc.). The parts of this section relating to required resources and expenses have some similarity to the budget section of the research proposal. However, the focus in the business plan is on the expectation of generating significant revenues over and above the investment and the ongoing expense.

The financial information for early stage high-tech startups is usually quite abbreviated and typically focuses on:
- Investment required — in resources and time — to reach key development milestones.
- The burn rate: The difference between the cash expenses and any revenues, i.e. the rate at which a start-up (or otherwise unprofitable company) is using up or "burning" the investment.
- Range of potential income upon success.

For later stage financing, pro-forma financial statements are expected. These statements indicate;
- How much money will flow into the company (from investments, loans, and hopefully ultimately sales of products and services),
- How much money will flow out of the company (for purchasing equipment and materials, paying salaries, rent, fees for various services, etc., and ultimately as dividends to investors).

Each of the items is calculated based on various assumptions and the assumptions are explained.

These details are usually organized in tables based on standard accounting formats. The three most important financial statements are described in the following paragraphs.

Balance Sheet. The balance sheet lists the end of year assets and liabilities of the company, starting with the most liquid. Assets include cash (i.e. bank deposits), securities, accounts receivable, and the value of owned real estate, buildings, and equipment. Liabilities include loans, accounts payable, and shareholders' equity. The statement of shareholders' equity may be shown in more detailed form separately. Shareholders' equity includes funds invested in the company by the investors and the equity built (or eroded) by the company's cumulative net profit (or loss). Two very important company assets, intellectual property and goodwill, are not normally included in the balance sheet, unless there was some cash transaction involving their purchase or sale.

Profit and Loss Statement. The profit and loss statement is a table of money flowing in and out of the company, by year, typically for a period of 5 years, or until the company is expected to be profitable, if longer.

Revenues, i.e. money flowing into the company, include in particular the sales of goods and services. These may be listed on separate lines for different products or classes of products. In the early years of a company, sales will typically be zero as products are not ready for market.

Expenses include the direct cost of the goods or services sold, e.g. raw materials, manufacturing, sales and distribution, etc. Other expenses include administrative overhead, interest expense, depreciation and amortization. Depreciation and amortization are the annual decrease of the value of buildings and equipment, respectively, with age. R&D is typically a major expense in a start-up company. The difference between revenues and expenses is the operating profit (or loss) of the company.

Cash Flow Statement. The cash flow statement is also a table of money flowing in and out of the company, similar to the Profit and Loss Statement, but considering only the components which are represented by cash. Cash inflow includes similar revenue items to those appearing in the Profit and Loss Statement although timing differences (e.g. when the customer actually pays for goods purchased) may influence the numbers. Additional cash inflow items arise from financing (e.g. receipt of a loan) and investments (e.g. increase in shareholders' equity). The expense side also differs. For example, depreciation is not included in the cash outflow but repayment of loan principal is. The difference between the cash expenses and the revenues is the "burn rate", i.e. the rate at which a start-up (or otherwise unprofitable company) is "burning money".

The financial plan includes projections based heavily on assumptions. For example, a key figure is how many units of the product will be sold at what price and when. It is critically important that these assumptions be carefully documented, because the company will want to revise its financial plan as real data becomes available, and expenses and revenues can be better predicted based on experience.

A common cause of failure of start-up companies is undercapitalization, i.e. insufficient investment to see the company through until it becomes profitable. Thus projecting the "burn rate" and securing the investment needed are very important.

While this section of the business plan is crucial, it is generally presented in a very brief form. Only the basic assumptions, the various financial tables, brief explanations of these tables, and the significant conclusions drawn from them (e.g. how much investment is needed, when will the company turn profitable, and what will the investors profit and when) are presented in the text of this section. More detailed explanations may be appended.

6.3.9 Appendices

The body of the business plan should be compact, typically 10–30 pages depending on the stage of development of the company and the investment amount solicited. More details may appear in one or more appendices with the expectation that only very interested parties will study them. These appendices may include CV's or résumés of key personnel, details of the technology, a complete market survey report, and details of the assumptions and models used in the financial plan.

6.4 Common Problems

A new technology does not necessarily create a business opportunity. A very common pitfall for scientists is the assumption that an innovative technology they have developed will be attractive in the marketplace. Business potential is measured by the value to the customer. However clever the technology, if it doesn't fill a need of the customer, it is not likely to be marketable. And if the market is too small to provide a good return on investment, investors will not invest.

A common problem in business plans is losing focus by dwelling on details rather than concentrating on the critical internal and external issues which will determine the success or failure of the proposed venture. The potential investor will be looking for these critical issues and how they are addressed, and will want to get through all subsidiary detail as quickly as possible. Writing long descriptions of technology and explanations of the financial plan are best avoided. It is preferable to keep these sections short and aimed at non-experts, and place the details in appendices.

Fulfilling the needs of the reader, i.e. the busy potential investor, means that the use of standard text ("boiler plate") in business plans should be completely avoided. Any generic statement that could be made for any business plan, i.e. that is not specific to your plan, has little value to the potential investor.

Different versions of the business plan may be prepared for different types of potential investors and often should be. Focusing on the specific concerns of venture capitalists, large corporations and government agencies is more easily accomplished in separate documents.

Another issue is internal consistency. The technology in the Products and Services section must meet a market need expressed in the Market section. The strategy expressed in the Production and Marketing Strategy must result in a product or service that is brought to an appropriate link in the production chain described in the Market section. The Operational Plan should express how to execute this strategy, and the management structure described in the Management and Organization section, as well as the company strengths expressed in the Company section, must be appropriate for managing and organizing this plan. Finally, the Financial Plan must provide a convincing basis for raising the funds necessary to execute this plan, and project an enticing profit in a reasonable period of time. Lack of internal consistency indicates either lack of care by the authors or worse, lack of understanding by the entrepreneurs of the information required by potential investors. In either case, an inconsistent plan can nullify the chances of obtaining funding.

6.5 Summary

The business plan serves as a planning tool for scientists and engineers to determine if some development in their lab might be profitably commercialized, and as the key document to convince potential investors to invest the necessary funds. The plan should be compactly written and concentrate on the key issues that will determine whether commercialization will be profitable.

For further reading:

Cynthia Kocialski, *Business Plans That Work*, eBook ISBN 9870104675-7288-0, 2013
John L. Nesheim, *High Tech Start Up: The Complete Handbook for Creating Successful New High Tech Companies*, revised, The Free Press, 2000.
William A. Sahlman, *How to Write a Great Business Plan,* Harvard Business Press, Boston, Massachusetts, 2008 (Originally published in the Harvard Business Review, July 1997)

Chapter 7

Patents

7.1 Introduction

Many researchers invent something with commercial potential in the course of their investigations, and wish to patent it. Although in principle the inventor can prepare and file a patent application on his own, considerable knowledge of patent laws and procedures is needed to do so effectively. Most patents are prepared and filed by professional patent agents or patent attorneys, on behalf of the inventors or their employers, and the inexperienced inventor is well advised to use their services. The intent of this chapter is to introduce researchers to patents, and how patent applications are structured in comparison to research reports. The objective is to enable the researcher to prepare the background material needed by the patent preparer more intelligently, and to facilitate efficient dialogue between the researcher and the preparer.

Many of the details of the patent process differ from country to country. This chapter tries to present the broad outline of typical practice and is generally based on U.S. procedure. It is oriented at the researcher, and as such, the overview presented comes at the expense of legal exactness.

7.1.1 *What is a Patent and why apply for one?*

A patent or "letters patent" is a legal document issued by a government granting the holder the right to prevent others from practicing a specified invention within its jurisdiction for a specified period of time. The patent is a legal embodiment of a "social contract" between an inventor and the society, in which society's representative, the government, grants the inventor monopolistic rights for a period of time, and gains information in return, in the form of disclosure of the invention. Society benefits from this arrangement in two ways. First, the disclosure of the invention, with sufficient detail to allow one normally skilled in the art to practice the invention, contributes knowledge to the public domain and may serve as the intellectual spur for further inventions. Second, by increasing the potential profit of the inventor, the patent system provides incentives to invent and to commercialize inventions, which may be useful to society and stimulate the economy.

The patent allows its owner to prevent others from practicing the invention. It does not grant him the right to practice the invention, because the practice of the new invention may require the use of an existing patented invention, and require permission by its owner. The government does not pay the inventor any sum of money, but just the opposite — it charges substantial application and maintenance fees. Nor does the government prosecute those who infringe on a patent or otherwise enforce the rights of the patent holder. The patent holder himself must discover any infringement, and enforce his rights himself, usually through the legal system.

Given the above, why should an inventor apply for a patent? By being able to prevent others from practicing the invention, the patent holder has a monopoly on practicing his invention for the period of the patent in the jurisdiction of the government granting the patent. Thus he or she may be able to profit substantially more from the invention than if others could also practice it. But for this monopoly to be worthwhile, the invention itself must be successful both technologically and commercially. Sufficient customers must be willing to buy the product or process protected by the patent for the invention to be profitable. A patent on a product or process that no one is willing to buy will not be profitable. It should be noted that the protection afforded by a patent is often very narrow, and disclosure of a commercially successful invention may spur further inventions which are outside the specific claims of the first patent, and hence not prohibited. In some cases, secrecy may be a better strategy for protecting an invention.

Patents serve an additional function — they impress potential investors. A patent indicates to them the seriousness of the inventor and the invention. And the investors believe that it increases the potential for profit, since the monopoly right can be potentially enforced.

> ## Other types of Intellectual property
>
> Patents are a type of intellectual property (IP), intangible property created by intellectual endeavor. A few other types of IP are described below.
>
> **Trade secrets** are processes, procedures, and information which are known only to a specific person or company. Generally, a company will seek to protect secrets that give it an economic advantage. It may do so with a combination of physical security and confidentiality agreements with its employees and others who are parties to the secret. However, if another party legally discovers the secret, e.g. through its own research, then the original secret holder has no legal means to prevent others from using this secret. A company generally weighs carefully when to keep inventions secret, and when to disclose them in a patent application.
>
> **Trademarks** are words or symbols which uniquely identify a product, or products, of a company, and service marks do the same for a service. In the U.S., both patents and trademarks are administered by the U.S. Patent and Trademark Office.
>
> **Copyright** is the legal right to copy or otherwise reproduce or perform literary, artistic, or musical material, or to permit others to do so. In the U.S. and most countries, the originator is automatically granted the copyright upon its origination. In the U.S., there is no need to register the copyright with the U.S. Copyright Office but there may be advantages in doing so. The requirements for journal paper authors to avoid violating the copyright of others and to assign their copyright to the journal publisher were discussed in sections 3.1.3 and 3.2.5.

7.1.2 Requirements for obtaining a patent

Patents are issued for novel, useful, effective inventions that have been reduced to practice. These elements are examined in the paragraphs below.

7.1.2.1 Invention

The invention must be a new product or process. In some countries, a product may include engineered living organisms, and processes may include medical procedures. Computer codes *per se* cannot be patented; however, they may be copyrighted. Furthermore, the process embodied by a computer code may be patented. For example, a novel digital signal processor might be controlled by an embedded computer code. The computer code itself cannot be patented, but it may be copyrighted. However, the signal processor might be patented, and furthermore the process of manipulating the signal might also be patented, provided that the signal processor and the process meet all of the requirements set forth in this chapter. Scientific discoveries and mathematical solutions *per se* cannot be patented, but an invention based on a new discovery or solution may be patented.

7.1.2.2 Reduced to practice

"Reduced to practice" means described in sufficient detail so that one normally skilled in the art can practice the invention without undue experimentation. Fulfilling this requirement does not require actually building the invention, though one can argue that it is impossible to prove that an invention is effective if it is not built and tested. Nonetheless "constructive" reduction to practice, i.e. submitting a patent application which describes the invention in sufficient detail for one normally skilled in the art to duplicate the invention, is the usual way in which this requirement is met. Cartoons in popular magazines of inventors sitting in the waiting room of the Patent Office with an array of strange inventions on their laps have no connection with modern reality.

7.1.2.3 Novelty

To be patented, the invention must be novel, i.e. new. Generally, to be considered novel, the invention must not have been publicly reduced to practice by anyone prior to filing the patent application, published by anyone in any form, nor have been offered for sale. Very few inventions are based on entirely new concepts. Most "inventions" are in fact improvements of existing inventions. If the improvement is novel, it can be patented.

One source of contention is the claim by several separate inventors to have invented the same invention. In this case, the question will be who invented it first. Until recently this was a contentious issue in the United States, which had a "first to invent" (i.e. first to reduce to practice) priority criterion. Inventors were required to maintain scrupulous records in the form of laboratory notebooks, in which new inventions were recorded, signed, and witnessed. Recently the United States has joined the rest of the world in adopting the "first to file" criterion, in which priority is given to the first inventor to file a patent application.

To obtain a patent, the initial public disclosure of the invention must be in the form of a patent application. If a researcher first publishes a scientific paper describing an invention, that invention becomes part of the public domain. The invention will thereafter not be novel, and neither the researcher nor anyone else will be able to obtain a patent on it. Even revealing the invention privately to another party, without an explicit notice of confidentiality, may jeopardize the novelty of the invention. There are a few exceptions that depend on the jurisdiction. Some countries have a grace period after some forms of public disclosure during which a patent application may be filed. However, the safe course for a researcher wishing to obtain a patent is to first file the patent application and only then publish.

7.1.2.4 Not obvious

The novel invention must not be obvious to one normally skilled in the art. To fulfill the "non-obvious" criterion, the invention must have at least one "inventive step" that is in some sense a breakthrough. The definition and determination of non-obviousness are the main issues of debate between patent applicants and patent examiners. Often patent examiners will look at the features of the claimed invention, and search for precedents for each one, and then claim that the invention is an obvious combination of prior art. This will put the applicant in the position of needing to demonstrate that this combination was not obvious.

7.1.2.5 *Utility*

The invention must be useful, i.e. it must perform some function. In practice, this criterion is not usually an impediment to obtaining a patent. It is joked that every mechanical device can be a paper weight, and every chemical invention either a fertilizer or a weed killer.

7.1.2.6 *Effectiveness*

Likewise, the invention must be effective. Also this criterion is usually not an impediment to obtaining a patent. However, a patent can be voided at some later stage in litigation if it is proven than an invention was not effective — i.e. it didn't work. But the value of a patent on an ineffective invention must be zero, so not much interest is shown in this criterion.

7.1.3 *Types of Patents and Patent Applications*

There are three basic types of patents: design, plant and utility. A design patent covers non-functional features that stand out to the consumer, and thus will rarely be of interest to the researcher. These are typically aesthetic or ornamental features, but might also include new fonts or icons which are incorporated in functional devices. A plant patent may be issued to an inventor who has asexually reproduced a distinct and new variety of plant, other than a tuber propagated plant or a plant found in an uncultivated state. The utility patent is the term used for useful inventions, other than plants, i.e. inventions that have utility. This is the most relevant type of patent for scientists and engineers.

Besides the regular patent application, the US has an optional *provisional* patent application. This application is less formal than the regular application, but must adequately describe the invention. It allows the inventor to quickly file, and establish a priority date, without requiring certain formalities in the application. However, priority is only established for what is in the application. Any new features added when submitting further applications will not benefit from the earlier priority date. A regular or a "PCT" (see the next paragraph) patent application must be filed within one year of the provisional application to maintain the priority date. Note that provisional applications are not published; the invention will only be effectively disclosed to the public if and when a regular or PCT application is published claiming priority from the provisional application. It should be noted that there is no "provisional **patent**", only a provisional patent **application**.

Most industrial countries belong to the *Patent Cooperation Treaty* (PCT). A PCT application may be filed in any member country by an inventor, or invention owner. The PCT application is examined by the search office of the applicant's choice, with the choice available varying based on the country in which the PCT application is filed. To obtain patent protection, however, the applicant must ultimately apply for a national patent in every jurisdiction in which he wants protection, within 30–32 months of the priority date, depending on the country. This is known as entering the national phase. Advantageously the PCT process gives the inventor or patent holder this period of time to examine the commercial prospects of the invention. A further advantage is receiving an International Search Report and Written Opinion regarding patentability before embarking on the relatively costly procedure of filing applications in each relevant national jurisdiction. Some jurisdictions (e.g., Argentina, Taiwan) are not members of the PCT, and applications must be filed in such countries within one year of the first patent filing.

Some jurisdictions have a type of patent for which the application procedure is less formal, such as the Inventor's Certificate in Russia and the Utility Model in Germany and China. These forms of patents may provide less protection than standard utility patents, and may be in force for a shorter period of time, however they typically require less novelty. Thus, they may be useful in a patent strategy which seeks to minimize patent costs in each jurisdiction for the degree of protection needed.

7.2 Structure of the Patent Application

Patent applications generally include Preliminary Information, Specification[1], Drawings, and Claims. The Claims precisely define the invention in terms of what aspects will be protected in the patent; the Specification and Drawings explain the invention, build the foundation for the claims, and build the case justifying issuance of a patent. In the paragraphs below, each section of the application will be examined in detail. Where relevant, the similarity to corresponding sections of the research report will be noted.

7.2.1 *Preliminary Information*

The first page of published U.S. patents contains the title of the patent, the inventors with their addresses, the assignee if any, an abstract, and citations. With the exception of the latter, this is similar to the first page of a journal paper. In contrast to journal papers, patent titles tend to be very short, not unique, and vague. There might thousands of patents with a title such as "coating device."

The inventors must include all those who contributed substantially to reducing the invention to practice, and only those. A substantial contribution does not include performing standard actions at the direction of someone else. Including persons who do not satisfy this criterion, or excluding someone who does, can jeopardize the legal status of the patent. Each inventor is listed with his residential address.

Under U.S. law, the default owner of the patent rights of an invention is the inventor. However, most technological employees sign an agreement with their employer, as a condition of their employment, in which they agree to assign all rights to any invention they created from their employment to the employer. Furthermore, some inventors who are not encumbered by such an agreement may wish at an early stage to assign the rights to their patent to a different individual or body, e.g. to their own company or to a company buying the invention. Such assignment can be

---

[1] Some texts define the specification as including the claims and drawings.

effected easily during the patent application process and will be noted on the published patent. Assigning the patent to a company simplifies raising capital for future research, development and marketing expenses, and controls the rights of the patent in accordance with well-established corporate government principles.

The patent Abstract is similar to the research report Abstract, and contains a short description of the invention, with a limit of 150 words. It, together with one of the drawings, will appear on the first page of the published patent, and give the reader a quick glance of what follows.

Citations in the patent are similar to those in research reports, and serve to establish the state of the prior art. Patents, however, tend to cite other patents, although they may cite the scientific literature as appropriate. They appear on the first page of the published patent. Patent writers often limit the number of citations to avoid admission of prior art, which must then be argued against. Typically, patent preparers include in the patent application only what they consider to be the closest prior art, and then only when they need the citation for the purpose of showing differences from prior art. However, all relevant prior patents and publications known to the inventor or the patent preparer are required to be submitted to the patent office on an Information Disclosure Statement.

7.2.2 Specification

The Specification section is usually the longest part of the patent application. It must fulfill the requirement of explaining the new invention in sufficient detail so that one skilled in the art can practice the invention. This is judged by a patent examiner, i.e. an official of the government patent office, whose sole job is to read and evaluate patent applications. In contrast to a journal paper, where we hope there will be a wide audience of readers, the entire effort of writing a patent (and particularly the specification) is initially directed to this one individual. The writing task is totally focused on convincing him or her that the described invention satisfies all the legal requirements for obtaining a patent from the government. While many others might read the published patent, their reading does not generally have any beneficial effect for the patent holder. However, in the event that the patent must be enforced through the courts, additional targets of the specification will include the judge and jury trying the case.

A particular language style has evolved in patent writing, which is neither pleasant nor easy to read, but which precisely conveys techno-legal meaning. The style includes jargon, some examples of which are noted in Table 7.1.

Table 7.1. Patent jargon

Jargon	Example	Meaning
Said	Coatings are deposited on a substrate having a first surface and a second surface. *Said* first surface of *said* substrate is exposed to *said* plasma jet.	≈ the, but more precisely indicates that the following object has already been mentioned previously.
Instant invention	According to one aspect of the *instant invention*, improved mechanical performance is obtained.	The invention described in the present patent application.
Comprising	....*comprising* the steps of:	Including (does not exclude additional items)
Aspects	Various *aspects* of the invention	A particular part or feature of the invention
Embodiment	In one *embodiment* of the invention, the fill is comprised of Ar and a small amount of Hg, as in prior art fluorescent lamps. In another preferred *embodiment*, the fill is air at reduced pressure.	A way of implementing the invention.
Preferred	In a *preferred* embodiment of the invention, microwave generator (10) is a conventional magnetron.	Specific (not necessarily preferable)

The specifications may be formally divided by headers into labeled subsections, or not. But in either event, the specification should contain the informational content described in the paragraphs below, and in this order.

7.2.2.1 Review of the Prior Art

The Review of the Prior Art fulfills the functions of stages 1 and 2 of the Introduction of a research report. It begins with a short broad background section, placing the invention in its broad technological context by briefly describing the field of the invention. For example, a patent for a novel improved vacuum cleaner might begin with a paragraph about the need to clean carpets, the types of materials which must be removed, and the challenges in doing so.

The section then continues with a review of previous inventions intended to fulfill the above described or similar functions, and inventions intended for other purposes that have similar features to the present invention. As in stage 2 of the research report Introduction, the various previous inventions may be grouped together chronologically, by approach, or by degree of similarity to the present invention. The objective in presenting this review is to show the patent examiner that the present invention is novel. Typically, in patent applications, other patents are cited in this review, but other published material may be cited as needed. This section, however, is not to be considered exhaustive, and only includes the minimal prior art needed to understand the problem to be solved, and to highlight the difference with the closest prior art. In contrast, all of the prior art discovered by the inventor in his search must be included in the Information Disclosure Statement and will ultimately appear in the list of citations which will appear on the first page of the patent.

7.2.2.2 Need for the Invention

The Need for the Invention, or Problem Statement, is an optional paragraph that is similar to the Gap statement (stage 3) of the research report Introduction. It may first summarize the state of the previous art in 1–2 sentences, and end in stating why a new invention is needed, generally by stating that a particular need is not addressed, or inadequately addressed (perhaps because of a low value of some performance parameter, or by high cost, or safety problems with prior art inventions). The purpose of this paragraph in a US patent application is merely to explain the motivation behind the new invention. In Europe, novelty is determined in relation to the Problem Statement, and thus in the event that a problem statement is not provided, the European patent office may invent one, which may not be to your advantage.

If included, this paragraph must be drafted with care. In some cases, a major part of the invention was the identification of the need for the invention by the inventors. In such cases the paragraph will be drafted using language which states that the inventors identified the need, rather than more general language from which it might be inferred that the need was well known. In other cases, the existence of a well-known need which was heretofore unfulfilled can help establish that an invention was not obvious.

7.2.2.3 Brief Description of the Invention

The Brief Description of the Invention describes what the invention does, and gives an overview of how it works. This section serves as an introduction to the more detailed description which will follow. It is somewhat analogous to the Statement of Purpose – stage 4 of the Introduction and the Preview – stage 5 of the Introduction in the research report. In the United States, typically, this section simply mirrors the independent claims. If the patent is to be filed outside of the United States, it will usually mirror all of the claims.

7.2.2.4 Brief Description of the Drawings

The Brief Description of the Drawings is more akin to a list of the drawings than a description. Each drawing is described in a single sentence, in the most general terms, e.g. *"Figure 4 is a frontal view of one aspect of the invention"*. It should be noted that while drawings are very common in physical science and engineering patents, and are usually needed to describe the invention, there is no specific requirement for drawings and some patent applications do not have any. However, each structural claim element, e.g. parts of the apparatus on which a claim is based, must appear in at least one drawing.

7.2.2.5 Detailed Description of the Invention

The Detailed Description of the Invention is the heart of the specifications, and must describe the invention in sufficient detail so that one normally skilled in the field could practice the invention. Furthermore, this section also lays the groundwork for the subsequent claims (see section 7.2.3). This section is analogous to Methodology section of a research report.

Usually this section is organized around the drawings, and describes them in detail. Most will be detailed schematic drawings of the invention. The purpose of these drawings is to explain the invention. Every principal feature of the diagram is numbered, and the description refers to these features by number.

7.2.2.6 Embodiments

Embodiments are examples of the invention, in which specific choices of design parameters are given. In some respects, they are analogous to the Results in a research paper, and some patent applications in fact present performance results of the presented embodiments. Generally, presenting a number of embodiments strengthens the patent. Often in the claims, some particular value or range of values of design parameters is claimed. Embodiments should be given which use these values or span the range of values, and preferably give some meaning to those value ranges.

7.2.2.7 Mechanism of Action

The Mechanism of Action is a theoretical explanation of how the invention works, i.e. an explanation of the underlying science. It can be useful, but it is not required. Understanding the fundamental science is not required, provided that empirically the invention is effective.

7.2.2.8 Disclaimer

Usually the specification ends with a Disclaimer stating that everything in the Specification was given to explain the principles of the invention, that these explanations do not limit the scope of the invention, and that one skilled in the art could conceive of additional embodiments which are still part of the invention.

7.2.3 Claims

The Claims precisely define the invention in terms of what someone other than the patent holder may not do, if the patent is granted. This section is analogous to the Conclusions of a research report. In many jurisdictions, it may not contain any new material – every claim must be firmly based on material presented and explained in the Specification. To ensure compliance with this requirement, some professional patent preparers simply copy the claims into the Specification, or into the summary of the invention. As a good practice, patent preparers typically first write the Claims, and if the claims are acceptable to the inventor, the Specification is then written which supports and enables the Claims.

There are two types of claims: independent and dependent. Independent claims do not refer to any other claim, and represent the broadest claim for the scope of the invention. Dependent claims add some additional limitation to an independent claim or another dependent claim, and thus have reduced scope.

Often an "onion strategy" is used in designing the claims. The patent holder may intend to manufacture a particular device, or range of devices, with particular design parameters. Dependent claims may be written to cover these devices, and future devices, in all of their particulars. However independent claims are written in an attempt to obtain the broadest possible protection. The dependent claims are ultimately needed in the event of a court action, where the court may be persuaded that the independent claim is too broad, and thus not novel in light of certain prior art. The dependent claims may still be found to be valid, and hopefully will be deemed by the court to have been infringed.

Each claim must "exclude" all prior art. In other words, each claim must describe an invention which is not previously known. Typically, independent claims begin with the function or application of the invention, known as the preamble, and then a list of features which define the invention. For example, *A device for exterminating mice [function] comprising an optical detector, a triggering mechanism, and a laser* [*features*]. The preamble, however, does not define the invention, unless it is required to add meaning to the claim. Thus if there were some other existing invention such as an anti-missile defense system, or an inspection system for a production line (i.e. different function specified in the preamble), that incorporated the same features, then this invention would not be novel, and a patent will not be granted. In other words, the features define the claim, not the function in the preamble. To obtain a patent, the inventor would need to add non-obvious features that limit the invention so as to exclude all previous inventions.

The claims are numbered sequentially, and the first is always an independent claim. Dependent claims can only refer to claims which precede it. Each claim grammatically has the form of an object of a single sentence, whose subject and verb are "I/We claim".

The claims must be composed with great care. On the one hand, they must exclude all prior art in order for the patent to be granted, and on the other hand, they should give the patent the broadest scope possible. Only the Claims define the scope of the protection against competition given by the patent. If some feature is described in minute detail in the Specification, but is not incorporated in the Claims, then the patent will not protect it, and the inventors will have gifted the knowledge of this feature to the public domain.

7.2.4 Drawings

Drawings are usually essential in explaining an invention, and are appended to the application text. While informal drawings are permitted for provisional patent applications, formal drawings using standard conventions and complying with specifications set by the various patent offices must be provided at some stage.

Each feature in a drawing described in the Specification must be uniquely numbered. The numbers do not have to be sequential, and often multiples of 5 or 10 are used. It is useful but not required to have some form of internal logic in assigning numbers. The same number may be used in different drawings to designate an identical feature.

7.3 Processing of a Patent Application

The complete patent application includes various forms and legal declarations, and the text and drawings described above. It must be submitted to the relevant patent office in the form which it specifies. Electronic submission is increasingly allowed and even encourged.

The patent office will assign an examiner who specializes in the general area of the patent topic. Usually the first step is to check the application for technical completeness. The patent office will notify the application filer of any decision or action it takes — such a notification is known as an office action. Each action requires a response by the applicant, such as the payment of a fee, supplying additional documentation, or taking some decision. Generally, a time limit for the response is stipulated, and failure to respond in time can result in additional fees, or abandonment of the application. The examiner will search the relevant literature, in particular previous patents, and try to discover previous inventions which are not excluded by the present patent application claims. The examiner will disallow part or all of the claims on the basis of the previous art or because the novel feature in the present application is obvious. In Europe, the public can also respond to a patent application, and point out previous art, in what is known as the opposition period.

If the whole patent application or specific claims are disallowed, the applicant can respond by: (1) abandoning claims in the patent or the entire application, (2) modifying the claims (generally narrowing the claims, perhaps by adding the limitation of one or more dependent claims into the respective independent claim, and modifying a dependent claim to be a more limited independent claim), or (3) rebutting the examiner's argument that some feature was previously described or is obvious. Such a rebuttal must be well supported, and written factually and respectfully. In the United States, the applicant or his preparer may request a telephone interview with the examiner. During such an interview the applicant can seek to understand the reasoning of the examiner, and explore ways to modify the application to meet the objections of the examiner. Statements made during such interviews are not binding, but they can help the applicant in successfully revising the application so that a patent will be issued. However, "fishing expeditions" (i.e. seeking an instant response on multiple alternative strategies) are not allowed in the interview.

If and when the examiner is satisfied that an application meets all of the requirements, then the patent will be allowed. The patent is published and issued, generally after payment of additional fees. Typically, patents are issued for 20 years from the application date, depending on jurisdiction. Additional and often substantial fees must be paid to maintain the patent in force at various times (e.g. after 3, 7 and 14 years in the U.S.).

7.4 Summary

The patent is a legal instrument which gives its owners the right to prevent others from practicing their invention for a stipulated time period, thus increasing their potential profits from commercializing the invention. The patent application must disclose the invention in the Specification in sufficient detail so that someone normally skilled in the art can practice the invention, and precisely define the invention in a way that excludes all preceding inventions in the Claims. Most parts of the patent application are analogous to parts of the research report, though they differ in style and language. To be patented, the invention must be novel and not obvious to a person normally skilled in the art, i.e. it must contain an innovative advancement over previous inventions.

For further reading:

Claus Ascheron and Angela Kickuth, *Make Your Mark in Science: Creativity, Presenting, Publishing and Patents — A Guide for Young Scientists*, John Wiley & Sons, 2005

J.M. Mueller, *An Introduction to Patent Law*, Second Edition, Aspen Publishers, New York, 2006.

Chapter 8

Reports in the Popular Media

8.1 Publicizing Science and Technology in the Popular Media

Although scientists and engineers communicate mostly with their colleagues, they do need to communicate occasionally with a wider audience as well. Researchers working for universities, government research labs and non-profit institutions are usually dependent on the public for their funding, either through governmental grants or contributions of donors. Effective communication of their achievements and ambitions can help build public support for governmental funding and encourage private donations. Institutional public relations (PR) departments are concerned with these issues for the institution as a whole and will usually organize such publicity efforts.

Commercial enterprises are likewise interested in communicating with the public, to enhance their reputation as R&D leaders, to build interest in new products which they offer for sale, and to encourage investment. Furthermore, popular publicity encourages students to enter relevant fields, and hence may eventually aid in recruitment of new staff.

Most popular publicity is organized and often drafted by PR specialists. This chapter aims at familiarizing researchers with the various genres, to facilitate effective communications with PR professionals, as well as to offer advice for situations where researchers may face popular media on their own.

8.2 Types of Popular Media Reports

Several of the most common types of popular media reports on science and technology are described below, in decreasing order of their resemblance to the research report described in chapter 2.

The semi-popular medium closest to the archival scientific journal is the **trade journal**. Trade journals are directed at specific industrial sectors and are supported by advertising. Most trade journals contain articles of general interest to their specific sector. These articles typically explain innovations and present underlying scientific principles in semi-popular form. They are often written by researchers employed by companies in the sector. The level of the paper should be aimed at engineers, technicians, and managers practicing in the sector.

Some **magazines**, such as *Scientific American*, aim to inform the general public about scientific or technological developments. Articles are often written by the researchers involved. The level and language should be appropriate for an intelligent non-scientist, i.e. someone who has only very general scientific training.

Many **newspapers and other periodicals** report on scientific and technological developments, often on a features page or section in their weekend edition. Most such articles are written by staff reporters or science editors, who usually have a journalism rather than a science background. They are sometimes based on press releases or interviews (often by telephone).

Occasionally, various media present **interviews** with researchers, particularly in the aftermath of some exceptional event such as a dramatic discovery or invention, or a disaster, or the receipt of an award. Interviews on exceptional events will often solicit opinions on matters (e.g. the cause of some disaster) which might only remotely touch the researcher's area.

Many researchers and research groups maintain **websites**, which are often aimed at their colleagues and prospective students. However, because of their accessibility, they may attract the attention of the general public, and in particular secondary school students.

8.3 Suggestions for Communicating with the General Public

This section offers general guidelines for communicating with the general public and suggestions for specific media.

8.3.1 *General guidelines*

The objective of communication with our colleagues and with the general public is the same, namely conveying information. However, the type of information conveyed and how it is presented differ significantly.

When we communicate with colleagues, we concentrate on the research, exclude the researcher, and highly value precision. Depending on the breadth of the journal, we expect that the readers have considerable background in the field, at least an undergraduate degree in the general area of the target journal, and share a precise technical vocabulary.

The general public cannot be expected to have this degree of prior training and background. They are, however, interested in a good story, and the very fact that the general public reads science and medicine columns testifies to their interest. They will be particularly interested in the human aspects of the story, including how the research results may affect them personally, or at least how they may affect other people or even humanity. For example, will the research result in a cure or treatment for a disease, will it lead to some particularly useful technological development, will it facilitate improvement of the environment, or will it give a boost to the economy?

Our challenge is to convey our research to the public as a compelling story. One way to do this is to frame the research as good narrative story. After all, we do that even in research papers. In literary terms, the literature review is the "set-up", the gap is the "call for action", often there is "drama" — e.g. competing theories about the phenomenon. Finally, there is a "resolution" in the results and conclusions.

Metaphors and analogies can give non-experts some indication of an abstract or complicated phenomena in terms with which they are familiar. For example, describe impedance matching in electronics as choosing the right gear in a bicycle according to the slope of a hill and the strength of the cyclist.

The popular press looks for "newsworthy" stories. Among the types of science stories which may be newsworthy are those which may help people in solving or understanding problems which they face: diseases, disasters, and frustrations with technology such as electricity blackouts or computers which freeze or crash; or have a novel, unexpected, or important result. The challenge for researchers in communicating with the general population is on the one hand to tell the story in a way that will be interesting and understandable to the public, while, on the other hand, to retain the trust of the scientific community by avoiding the temptation to exaggerate or over-simplify.

The organization of the various genres for communicating with the general public mentioned in section 8.2 are very different, and detailed guidance is beyond the scope of this book. A few suggestions for the various genres are given below. The most important suggestion for preparing any popular publication is to read many examples from the target publication and to note the acceptable content and style.

8.3.2 *Articles in trade journals*

Although the motivation for writing in trade journals may be promotion of an organization or its products, the writer should keep the contents as objective as possible. Objective comparisons of products or technologies will bring credit and trust to the author's company or institution. Description of the advantages and disadvantages of new techniques and devices compared to conventional practice are very useful and most powerful if they are quantitative. Derivations of equations should be avoided. This type of article should focus on the technology and only mention the researchers if there is some exceptional value to their story. Jargon and abbreviations should be avoided, and if used, defined clearly.

8.3.3 *Articles in popular scientific magazines*

Articles written by researchers for popular scientific magazines are an excellent way to communicate directly to the interested general public. Such articles should focus on the "research story": what was known or not known on the topic prior to the research, what was the objective of the research and why reaching this objective was important, what was done, what was found, and how the result was significant. These are the same organizational elements that appear in a research report, but the style will be very different. No advanced training or background should be expected in the readership. Therefore, everything must be presented in terms that the general reader will understand. Everything must be explained. Only ordinary terms (i.e. no jargon) should be used. Given the need to explain and space limitations, such articles should concentrate on the big picture – there will not be sufficient room to describe small details. In contrast to research reports, the researcher's motivation, personal path to the discoveries or results, and the relationships between the various researchers and how they met, might interest to the reader, especially if there was something unusual about the circumstances. The main focus, however, should be on the research results, and their implications to humankind.

8.3.4 Newspaper articles and press releases

Science and technology articles in newspapers are nominally written by science reporters on the newspaper or news service, or by stringers or freelancers associated with them. Many such articles originate as press releases issued by organizational PR personnel. Accordingly, such press releases should be in the form of an article in the target publications. Small newspapers may publish such stories with only minor editing. Serious reporters and news services will use press releases as a starting point for their own journalistic research. The impetus for a story may also come from journal articles and presentations made at scientific conferences. The reporter may interview the researchers involved and likely other researchers in the field from other institutions. He or she may request additional information to verify the validity of the new findings, such as justification of the statistical significance. Neither the researcher nor the PR person will have any control on how the press release or interview is used or on the content of the published story.

Whereas in research reports to colleagues the emphasis is on providing sufficient information to enable the reproduction of your results, in articles in the popular press the emphasis should be on clarity for the lay reader. These articles should present the big picture of the field of research and the research results, and avoid technical detail. The significance of the result or the event, for the local community, the nation, or humankind should be explained in layman's language.

8.3.5 *Interviews*

Occasionally reporters from the general print or electronic media interview researchers. Interviews may be part of a wider story or may be the focus of a story. Often researchers from another institution are asked to supply "outside comment" on new discoveries, e.g. to state how the discovery is viewed by the scientific community, or to verify that the claims are not exaggerated.

Before agreeing to participate in an interview, it is recommended that the potential interviewee read, listen, or watch other interviews conducted by the same interviewer or in the same program. Keep in mind that in this form of publication, the interviewee has almost no control of what is aired or published. Accordingly, researchers should weigh carefully whether to accept or decline an interview invitation. If the researcher accepts, the research on the interviewer and the program should be used to prepare for the interview.

In speaking to the general public, no matter in which media, researchers must overcome their tendency to qualify each statement, and to provide too much detail. Get straight to point, and concentrate on the big picture rather than the details. In radio and television interviews, plan each answer to fit into a "sound bite" of 10–30 seconds duration. If possible, first answer direct questions directly and quickly. Essential qualifications should fit within that initial sound bite. Further qualifications and explanations may then follow in subsequent sentences.

Practice telling your story with people you know both outside your academic circle and outside academia. Revise the story according to their feedback.

> **Practice Communicating your Research**
>
> Aunt Freda always was keenly interested in what I, Raymond Boxman, was doing, as a child, as a student, and as a young engineer/researcher. She was a bright, educated woman but she had only minimal science education, and none in my field. Nonetheless, she demanded an explanation of what I was doing, in terms she could understand. She wanted to know the basic physics involved and why my research topic was important. To this day, the lessons I learned from communicating science with Aunt Freda guide me when I speak to non-specialists in my field.
>
> You might not need to search very far for an "Aunt Freda", a relative or friend who is not a scientist or engineer but who is interested in what you are doing. Use your personal "Aunt Freda" to develop the story of your research in terms that a lay person can understand. In writing material for the general public, ask yourself after each sentence, "Would Aunt Freda understand up to here?" In preparing for an interview with the general media, practice your story with your personal "Aunt Freda". As you do so, you will find it easier to express clearly to a general audience what you are doing and why.

In an interview with an experienced science writer in the written media, the researcher's task is to explain his research in terms that the interviewer will understand. A little research into the interviewer's background and previous articles will help in this task, as will good dialogue between the researcher and the interviewer. The interviewer's task when writing the article is to explain the research to the public in terms that the public understands. The interviewee may request at the beginning of the interview that the interviewer read back direct quotes, in order to check for accuracy, but not all interviewers will agree.

In contrast, in a radio or television interview, the researcher must deliver the message directly to the public. There is little opportunity for feedback. Thus it is extremely important to tell the story in short sound bites in a way that the public will understand but without being condescending.

Television presents the greatest opportunity in reaching a wide audience, but also the greatest challenge. While statements must be "sound bite" short and simple, they must be fully truthful. During the interview, look directly at the interviewer and do not be distracted by cameras, monitors, or other staff. Avoid fidgeting in the seat, or making excessive hand gestures. For best results, practice delivering the message, and answering questions, including difficult and aggressive ones. Arriving at the studio early (~30 min) allows time to become familiar with the set-up and the staff, and to take a few minutes to compose oneself. How the researcher's message is delivered, including both visual and vocal aspects, greatly influences how the message is perceived by the television audience.

8.3.6 *Websites*

Most researchers or their research groups maintain web pages within their institution's website. These web pages are usually aimed at colleagues and potential graduate students. Typically, such sites contain: a home page giving the 'big picture", a CV page, a publications page, and a research projects page. Often the sites include links to other groups, other pages dealing with their field, and conferences and publications in their field.

However, given the efficiency of web search engines, anyone, including members of the general public with an interest in the topic, is apt to access such web pages. Hence this is a golden opportunity to present material to the general public in a manner which is completely controlled by the researcher.

Accordingly, include material which targets a general, non-specialist audience. Consider aiming the text at the level of high school students, who often search for material for school projects. Ideally, the web pages will tell the story of your research, so that high school students and laymen can understand it. The site should be rich with visuals: photographs and video clips illustrating some aspect of the research. More interest may be generated if the page is designed so that fast loading text is loaded first, and slower photographic material afterwards. It is usually best to start with the "big picture", defining and describing the field in general, and only then to present some specifics of the researcher's or group's activities. And if possible, relate the work to everyday experience or the interest of the general reader.

For further reading:

Daniel Q. Haney, "A Reporter's Advice to Medical Researchers", *Clin Cancer Res* 2005; 11(19) October 1, 2005.

Kathleen A. Kendall-Tackett, *How to Write for a General Audience: A Guide for Academics Who Want to Share Their Knowledge with the World and Have Fun Doing It*, American Psychological Association, Washington, D.C., 2007

Chapter 9

Correspondence and Job-Hunting

9.1 Business Letters

The business letter is a form of correspondence which is short and to the point. Typically, a researcher needs to write such letters to request information or price quotations from vendors of supplies and equipment, as part of the paper submission and review process, and while looking for a job.

The business letter contains the following elements:
- Letterhead
- Date
- Reference Numbers (recommended but optional)
- Return Address (both physical address for material delivery and electronic for correspondence)
- Subject
- Salutation
- Body
- Signature

These will be examined in the following paragraphs, and an example is given in Fig. 9.1.

9.1.1 *Letterhead*

The letterhead is generally generated by the company or institution and may be printed on paper for official correspondence. More commonly, it can be incorporated in the header of a word processor template for the first page of a letter, especially if a color printer is used. It contains the name of the company or institution, its logo, and all relevant communications data, e.g. (snail) mail address, e-mail, telephone and fax numbers, and home web page. In some cases, the writer, the writer's title, and department are also incorporated into the letterhead. The letterhead only appears on the first page of the business letter.

9.1.2 Date

The date the letter is sent is placed below the letterhead, conventionally at the right margin.

9.1.3 Reference numbers

Some organizations, e.g. often military and governmental bodies, assign a unique number to all outgoing correspondence. When responding to such a numbered letter, the sender's reference number should be noted.

9.1.4 Inside address

The name, optionally the work title or position, and address of the addressee is included in the letter, just as it would appear on the envelope. The name is written in the following form: title (Mr., Ms., Dr., Prof.), first name or first initial, middle initial or initials, if known, and family name (for example, *Prof. R.L. Boxman* or *Ms. Edith Boxman*). Including a work title (e.g. *Manager, Quality Control Department*, or *Dean, Faculty of Engineering*) is useful, in case the named individual has moved to a new position — whoever handles the incoming mail would decide if the letter should be forwarded to the individual, or sent to the person currently having that title. The address should be sufficiently complete to ensure delivery.

9.1.5 Subject

Each letter should have a concise subject, less than one line in length if possible.

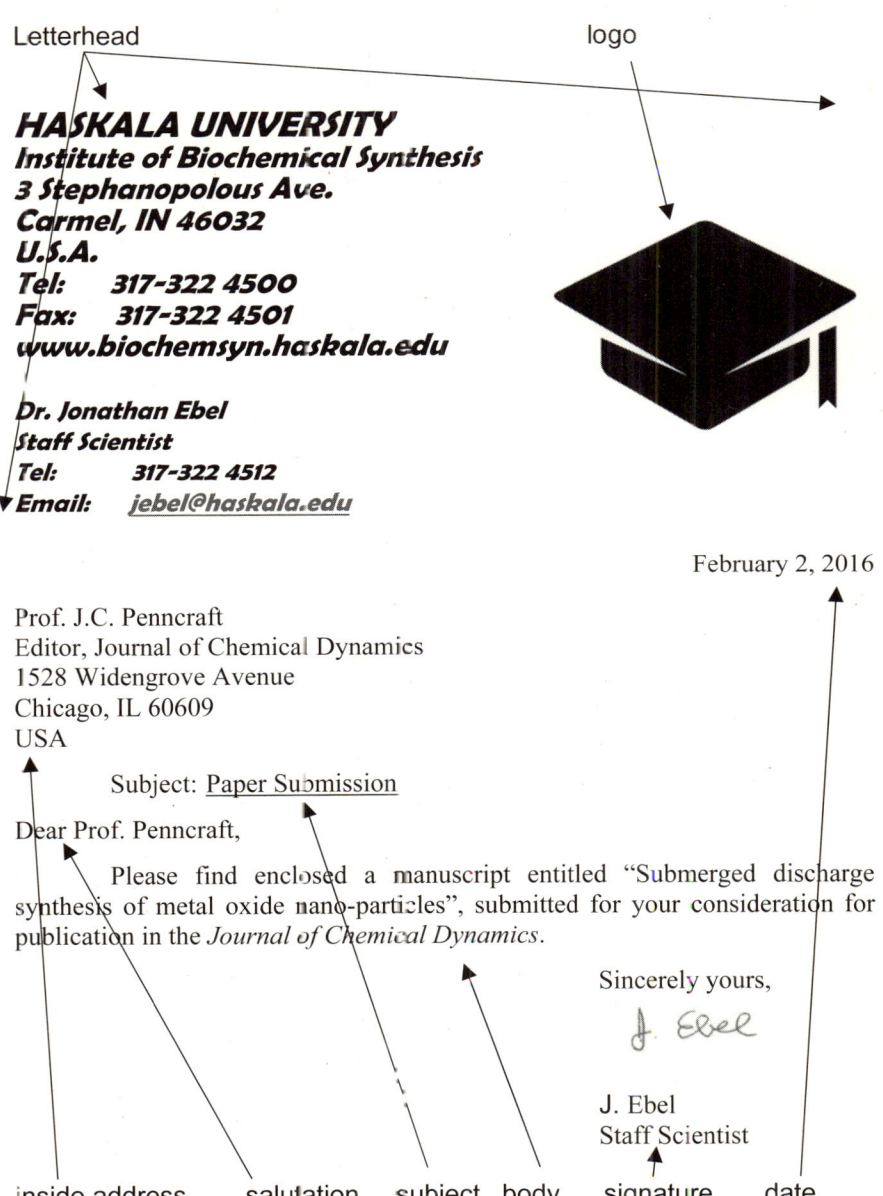

Figure 9.1. Example of a business letter.

9.1.6 Salutation

There are two acceptable forms of salutation. The conventional form is "Dear title family name" (e.g. *Dear Dr. Boxman*). Only in cases where the writer and the addressee use first names when speaking with each other is the form "Dear first name" (e.g. *Dear Ray*) acceptable. "Dear full name" (e.g. *Dear Raymond L. Boxman*) and "Dear title first name family name" (e.g. *Dear Prof. Raymond Boxman*) are not acceptable forms.

9.1.7 Body

The body of the letter is short and to the point. The cover letter for a paper submission would be a single sentence (e.g. *Please find enclosed a paper entitled "Prairie Beaver Population during Global Warming", submitted for your consideration for publication in North American Zoology.*)

An inquiry requesting information or a price quotation should identify what you want (e.g. specification, descriptive brochure, price quotation), and identify the product, either generically with specifications, or by model number. It is useful also to add a few sentences about your requirements and intended application. This will allow the addressee to suggest alternatives.

A business letter applying for a job might be longer, up to two pages in length. In it, the applicant would make the best argument possible for his or her candidacy. A job cover letter should be to the point, demonstrating that the candidate fits the position and the target institution in terms of experience and future plans.

In general, business letters should be confined to two pages in length. If more space is needed, usually the business letter serves as a cover letter for a longer document, which might be organized as a report with section and sub-section headings.

9.1.8 Signature

The signature is comprised of "Sincerely yours", the sender's handwritten signature, printed name (first name, middle initial, if used, and family name) and work title. There are alternatives to "Sincerely yours" (e.g. yours sincerely, yours faithfully, very truly yours) for particular types of letters and relationships between the writer and the addressee, but "Sincerely yours" is always acceptable.

9.2 Electronic Correspondence: Fax and E-mail

An increasingly large fraction of professional and business correspondence is now transmitted electronically, so much so that conventional "snail" mail is almost obsolete, except for legal documents requiring an original signature. At present, there are two principal forms of professional and business electronic communications: fax and e-mail. However, the use of fax is declining, and other forms of electronic communication will undoubtedly evolve.

Fax can be considered simply as another means of physically delivering a business letter. The same conventions described in Section 9.1 may be followed, with one exception: The fax number should be included in the form "*sent by fax to +1 860 423 2233*", either following the full "snail" mail address, or instead of it.

E-mail is exceedingly convenient for professional and business correspondence, as well as personal correspondence. The comments which follow are confined to business and professional e-mail. The main principle of the business letter should be followed — keep it short and to the point. If lengthy information must be transmitted, formulate it as an attached document. The body of the e-mail in this case serves as a cover letter for the longer attached document. This cover e-mail, however, should summarize and deliver the "bottom line", i.e. the conclusion, of the attached document. This is for the convenience of the reader, who can then decide when and how to deal with the long document, and where to store it in the meantime.

The structure of e-mail clearly identifies the address of the sender and the recipient, and provides a date and a subject line. Thus there is no need for duplicating this information in the body of the e-mail. Care should be taken that the subject line is accurate. The all too common tendency to use "reply" to some irrelevant message as a short-cut for finding the recipient's e-mail address usually results in an incorrect subject line. If the new correspondence is on a different subject, an appropriate new subject should be entered, and the string of irrelevant prior messages should be removed.

The body may begin with the salutation, following the business letter guidelines in Section 9.1.6. It is then followed by what is in the body of a business letter as described in section 9.1.7.

The current convention for the signature in e-mail is to write "Regards" (rather than "Sincerely yours"). Our recommendation is that the full name and job title should follow, even though the name of the sender might be apparent from the e-mail address. Likewise, we recommend that under the signature, the other communications information of the sender should be provided, e.g. phone and fax numbers, snail mail address, and professional web page. In most e-mail programs, insertion of this information can be automated by preparing a signature file.

9.3 Résumés, Curricula Vitae, and other Job Hunting Documents

Résumés and curricula vitae (CVs) succinctly summarize a person's education, qualifications, and experience. They are used in job hunting, as well as other personnel activities (e.g. promotion), and are often incorporated into personal or institutional websites. These documents are similar in style and contain many common elements. Résumés are used in the business and industrial communities, while CVs are used in academia.

Both documents are written so that the reader can get an impression of the subject's background in a glance. Information is conveyed in "bullet point" style (similar to lecture slides). Complete sentences are avoided.

The following items are standard in both:
- Full name
- Date
- Communications details: telephone numbers, address, email, etc.

- Immigration status, if you are applying for a position in a country where you are not a native-born citizen.
- Education (starting with the most recent, ending at college/university level for researchers): dates, institution, major, degree.
- Military experience (if relevant): dates, military branch, final rank, significant assignments. Include only if professionally relevant or if needed to show the continuity of your experience, and if appropriate to the culture of the target organization.
- Work experience (starting with the most recent): dates, employer, job title, significant assignments. In some fields, significant university level teaching experience is detailed separately immediately following Education.
- Honors, awards, and distinctions (if any): dates, granting organization, name of honor, short explanation (a few words).
- Professional certifications or licenses (if any — e.g. teaching certificate, professional engineer's license).
- Professional activities (if any — e.g. leadership or organizational roles in professional societies, editorial position on a journal, or other volunteer activities which advance the profession or public awareness thereof): dates, activity, and brief explanation if not obvious.
- Languages: language, degree of proficiency.

Personal Data

Until recently it was standard to also include personal data: date of birth, and marital and parental status. This is no longer given nor requested in the U.S.A, to comply with anti-discrimination policies.

9.3.1 Résumés — additional items

Résumés for young professionals are typically one to two pages, but might be longer for mid-career professionals depending on their employment record. In addition to the elements listed above, the résumé should include the following items:

- Objective: a short statement (one to two lines) about the job sought. This may be composed to include an immediate objective for the

specific job and an eventual career objective. Many job-seekers prepare slightly different objectives for each type of job sought (e.g. an academic research position, an academic position with a heavy teaching component, an industrial position) and then further refine the objective for specific job openings. The objective is placed near the beginning of the résumé, immediately following your name.
- Communal activities, hobbies, interests: this sort of information rounds out the more standard information noted above and may serve as a personal bridge to the prospective manager if there is some commonality of interest. This section might include volunteer work in charitable, youth, religious, social, communal or political organizations, noting leadership or organizational roles and a brief statement of favored leisure activities.

9.3.2 *Curricula Vitae — additional items*

CVs may be as long as necessary to include all of the relevant information, but are in the same telegraphic style as used in the résumé. Job objectives and communal activities are not included.

Some researchers include membership in professional societies and attendance at conferences. Such information does not convey any information other than that the person or his organization paid for registration fees or dues and should be omitted.

However, a list of publications is either added or appended as a separate document. If added, the publications should begin on a new page. The list should be separated into the following categories:
- Refereed Journal Papers: authors (in the order they appear in the paper), title, journal, volume, inclusive page numbers, year (and electronic reference number, if relevant).
- Conference Presentations: authors (in the order they appear in the presentation), title, name of the conference, place, date. Note if the presentation had some special status (i.e. invited or keynote talk).
- Chapters in Books: authors (in the order they appear), chapter title, and publication details of the book (overall author or editor, book title, publisher, city, year).

- Books: authors (in the order they appear), title, publisher, city, year.
- Patents: inventors (in the order they appear), title, country, patent number, issue date.
- Theses: title, degree, institution, year.

Other categories (e.g. articles in popular media, unrefereed papers posted to websites, etc.) may be added as relevant. A uniform style should be used for publications in each category. In cases where a paper is listed but hasn't yet been published, its status should be accurately noted (accepted, submitted, in preparation). Note that some institutions ignore papers in preparation.

A sample résumé is presented in Table 9.1, and a sample CV in Table 9.2. The differences between résumés and CVs are summarized in Table 9.3.

Table 9.1. Sample résumé.

JOSEPH R. RASMUTH, Ph.D.

Résumé

March 2015

Address	Department of Materials Science and Engineering, Midwest University, 52 Technology Blvd, Chicago, IL, 60201, USA
Residence	1220 Washington Ave., Apt 3, Chicago, IL 60610, USA
Telephone	+1 847 471 4339 (office); +1 312 244 9864 (cell)
e-mail	jorrasmuth@gmail.com
Status in the USA:	US Citizen

GOAL: to utilize my experience and education in materials science and electronics in an industrial research environment which will allow professional growth into an R&D leadership position.

EDUCATION:

June 2015   Ph.D., in Materials Science and Engineering, Midwest University, Chicago, IL, USA.
Thesis Title: Super-alloy Synthesis Based on the Sc-Al-Ni System
Thesis Supervisor: Prof. A.N. Untkraft

June 2012   M.Sc. in Materials Science and Engineering, Midwest University, Chicago, IL, USA.
Thesis: Investigation of Creep in Sc-Al Alloys
Thesis Supervisor: Prof. A.N. Untkraft

June 2009   B.Sc. in Electrical Engineering, India Institute of Technology, Delhi, India.

EXPERIENCE:

September 2012 to present
BLICA_URE, Inc. Northbridge, IL, USA. *Consultant*- Developed simulation program for predicting mechanical properties of alloys subjected to mechanical and thermal stress. Advised in selection of materials for advanced aerospace applications.

September 2010 to June 2012
Midwest University, Chicago, IL, USA. *Teaching Assistant* — Conducted recitation classes and graded homework and exams in Introduction to Materials Engineering; conducted laboratory in Electronic Instrumentation for Material Engineers.

June 2009 to August 2010
Birla Metal Works, Bangalore, India. *Quality Control Engineer* — Developed an electronic system for acceptance testing of aluminum alloy rods.

PROFESSIONAL AND UNIVERSITY ACTIVITIES:

2013-2015
Vice-president, student branch of the Society of Materials Testing, Midwest University, Chicago, IL, USA

2008-2009
Membership chairman, student branch of the IEEE, IIT, Bangalore

HONORS AND ACCOMPLISHMENTS:

February 2012
Outstanding Teaching Assistant Award, School of Engineering, Midwest University, Chicago, IL, USA

April 2009
Elected to Eta Kappa Nu honorary society.

February 2009
1st prize, E.E. senior project competition, IIT, Bangalore, India

To date
Published 3 refereed journal papers, and issued one patent.

ADDITIONAL PROFESSIONAL SKILLS

Languages	English — fluent, Hindi — fluent; German — basic reading knowledge
Computer skills	**Programming Languages**: Python, Java, C, C++ **Operating Systems**: LINUX, Solaris and Windows **Computer Software and Engineering Programs**: MS Office, Maple, MathCAD, Matlab, CrystalMaker, Ansys, Thermo-Calc
Laboratory Equipment	Experienced in working with: MTS Testing Machine, Data Acquisition Systems, Stress-Strain/Load-Displacement Analysis Instruments

COMMUNITY ACTIVITIES AND INTERESTS

2010–2014	Student Alliance for Science Education — tutored Chicago inner-city youth in math and science
2012–2014	Indian Students Association — welfare committee chairman
2011–2012	Campus-wide Music Concert organization, attended by 1,000 people, Co-Organizer 03/2012 Cultural Fest'12 at Midwest University, Co-Organizer 05/2011
Interests	Hobby: model aircraft design Leisure: listening to jazz and folk music

Table 9.2. Sample CV.

JOSEPH R. RASMUTH, Ph.D.

CURRICULUM VITAE

March 2015

Address	Department of Materials Science and Engineering, Midwest University, 52 Technology Blvd, Chicago, IL, 60201, USA
Residence	1220 Washington Ave., Apt 3, Chicago, IL 60626, USA
Telephone	+1 847 471 4339 (office); +1 312 244 9864 (cell)
e-mail	jorrasmuth@gmail.com
Status in the USA:	US Citizen

EDUCATION:

June 2015	Ph.D., in Materials Science and Engineering, Midwest University, Chicago, IL, USA. Thesis Title: Super-alloy Synthesis Based on the Sc-Al-Ni System Thesis Supervisor: Prof. A.N. Untkraft
June 2012	M.Sc. in Materials Science and Engineering, Midwest University, Chicago, IL, USA. Thesis: Investigation of Creep in Sc-Al Alloys Thesis Supervisor: Prof. A.N. Untkraft
June 2009	B.Sc. in Electrical Engineering, India Institute of Technology, Delhi, India. Elected to Eta Kappa Nu honorary society.

EXPERIENCE:

September 2012 to present — BLICALURE, Inc. Northbridge, IL, USA. *Consultant*-Developed simulation program for predicting mechanical properties of alloys subjected to mechanical and thermal stress. Advised in selection of materials for advanced aerospace applications.

September 2010 to June 2012 — Midwest University, Chicago, IL, USA. *Teaching Assistant* — Conducted recitation classes and graded homework and exams in Introduction to Materials Engineering; conducted laboratory in Electronic Instrumentation for Material Engineers.

June 2009 to August 2010 — Birla Metal Works, Bangalore, India. *Quality Control Engineer* — Developed an electronic system for acceptance testing of aluminum alloy rods.

PROFESSIONAL AND UNIVERSITY ACTIVITIES:

2013-2015 — Vice-president, student branch of the Society of Materials Testing, Midwest University, Chicago, IL, USA

2008-2009 — Membership chairman, student branch of the IEEE, IIT, Bangalore

HONORS:

September 2014 — 2nd prize, best student presentation, Materials Research Society Fall Meeting, San Jose, CA

February 2012 — Outstanding Teaching Assistant Award, School of Engineering, Midwest University, Chicago, IL, USA

JOSEPH R. RASMUTH
LIST OF PUBLICATIONS

Articles in Journals

1. A.N. Unstkraft and J.R. Rasmuth, "Creep in Sc-Al Alloys under Thermal and Mechanical Stress" *Aerospace Materials*, Vol. 22, pp. 23–31, 2011.

2. A.N. Unstkraft and J.R. Rasmuth, "Influence of Ni additions on Sc-Al Alloy Creep Performance" *Aerospace Materials*, Vol. 23, pp. 223–228, 2013.

3. J.R. Rasmuth, X.H. Chen, and A.N. Unstkraft, "Precipitation Hardening in Ni-Sc-Al Alloys" *Acta Materia*, Vol. 48, pp. 2338–2345, 2014.

Conference Papers

1. A.N. Unstkraft and J.R. Rasmuth, "High Temperature Superelasticity in Ni-Sc-Al Alloys under Thermal and Mechanical Stress" Materials Research Society Fall Meeting, San Jose, CA. 24–25 September 2014.

Patents (issued)

1. J.R. Rasmuth, "Electronic Rod Tester", U.S. patent 8,581,142, May 25, 2024.

Table 9.3. Résumés and CVs — comparison.

	Résumé	**CV**
Target	Business and industrial companies	Academic institutions
Length	Short (≤2 page for young applicant)	As long as necessary
Additional Items	Personal objective	List of publications
	Communal activities, interests, and hobbies	

9.3.3 *Additional documents for tenure track academic positions*

Tenure track academic positions in leading universities are highly competitive. These institutions seek exceptional candidates who will be research leaders in their fields. In addition to a CV and list of publications, they typically request additional documentation to aid in identifying a short list of exceptional candidates. The form of the documentation may vary from institution to institution, but will typically include the following elements:

- Teaching statement:
    o Brief listing of prior teaching experience
    o List of existing courses of the target academic unit which the candidate feels competent to teach
    o New courses (typically at an advanced level) which the candidate would like to offer
- Research statement
    o Very brief summary of research conducted by the candidate, and its significance
    o 5-year Research Plan, explaining what the candidate plans to investigate during the next 5 years, **and its significance**
    o Resources that will be needed to execute the plan
        - What will be needed from the institution
            - Office and/or laboratory space
            - Money to buy equipment
            - Operating expenses
        - What may be expected to be obtained from outside sources, and what sources are in principle possible.
- Professional development plan
    o Professional goals during the coming 5 years
    o Planned steps to reach these goals

In all of the above, the emphasis is on the future. Past activities and accomplishments should be summarized as briefly as possible, to the extent necessary to understand the future plans. Leading universities seek new appointees who have both vision and the necessary skills to realize this vision. The objective of the above documents is to demonstrate to the institution that the candidate has both.

9.4 Some Suggestions for Job Hunting

9.4.1 *Career planning*

The documents described in section 9.3 should be prepared to present the accomplishments and skills of a job candidate in the best possible light. But first of all, there must be accomplishments which will interest the prospective employer. It is valuable to consider your future career direction at as early a stage as possible and plan concrete steps that will further your career. For example, someone desiring an academic career should acquire frontal teaching experience and have a good publication record, including significant publications in which the candidate is the sole or lead author. Someone aiming at an industrial career should try to acquire industrial experience through co-op assignments, summer internships and consulting work. More detailed advice is offered in the texts listed at the end of the chapter.

In spite of what is stated above, a wise young investigator should be prepared to seize opportunities that present themselves. Not all successful careers proceed according to early plans.

9.4.2 *Goals, realistic expectations and viable alternatives*

The first step in job-hunting is deciding on your goals, and translating them into an ideal job description. Is your inclination towards research, development, operations, or business? Are you looking for an industrial or an academic position? Are there geographic considerations? Do you want to work in some particular narrow field, or are you flexible in terms of subject? The answers to these questions will help describe your ideal job, and guide your search for this job.

However, you should also realistically evaluate your chances of landing this ideal job. For example, tenure track positions in urban universities are in high demand. You should try to compare your record against those who recently obtained such positions. It is good practice to have a "fallback" alternative.

9.4.3 Research prospective employers

Before contacting prospective employers, use your research skills to investigate the various options, with two objectives: to filter the list of alternatives by ranking them according to whatever criteria are important to you and to obtain the requisite information to make as positive an impression as possible. In the case of companies, you will want to learn about their products and services, their customers, a little about their history, and as much as you can about their plans and ambitions. In addition to the usual web-based resources, public companies file annual reports which are generally available. If you are applying to a specific department, try to learn as much as possible about it by searching for papers and patents written by key personnel.

Pay attention to the language used by the company. Familiarity with the terms and style the company uses to describe itself and its goals can help you tailor your job objective on your résumé. Moreover, it can provide cues to aid you in interviewing for a job in this environment.

9.4.4 Use contacts

Landing a good job depends not only on what you know, but also who you know. R&D (and other) managers consider good employees very important for the success of their company or institution, as well as their own personal success, and taking a chance on an unknown candidate is considered very risky. Accordingly, it is good practice to become personally acquainted with potential bosses. Managers relate much more positively to applications from those whom they know than to complete strangers.

Ph.D. students and post-docs can develop the direct acquaintance of those who have influence on hiring by participating at conferences and by corresponding on scientific issues. Furthermore, indirect contacts can be developed through your thesis supervisor, other senior professors in your department, and recent graduates from your group. You should not be reticent in politely asking for their assistance.

9.4.5 *Tailor your pitch for the specific job*

After selecting potential employers according to your goals and your fallback strategy, researching them, and, if possible, obtaining an introduction from a mutual acquaintance, you will need to formally apply for a position. The first formal contact is usually in the form of a business letter, accompanied by your CV or résumé. The letter should make the best case possible for why you, the applicant, are best suited for the specific job which is offered, or in general for a job in that particular company or institution, and even more, in the particular department. This letter should utilize what you learned in your research of the employer. To the extent possible, your letter should point out what aspects of your training, experience, or research would assist the organization in furthering its specific goals or missions, and at the least, you should explain your specific interest in the job or organization. Enthusiastic superlatives about interest in the job should be avoided. Examples of poor and good sentences in a job application letter are given in Table 9.4. Furthermore, you should consider tailoring your CV or résumé to emphasize those aspects of your background that are most relevant to the potential employer.

Most employers receive many more applications than they have available positions. Some will request additional information, such as a completed application form, from all applicants. Others will reply only to selected applicants and may request additional information from them. Post-doc positions may be decided on the basis of this information and possibly following a tele-interview (e.g. via telephone or internet). Hiring for higher-level positions will probably involve an in-person interview.

Table 9.4. Examples of poor and good sentences in job application letter.

Poor	Good	Why
I am applying to your company because I am eager to work for the market leader in personal electronics.	I am very interested in personal electronics, and chose elective courses accordingly. My undergraduate project was designing the sensing circuit for a proximity activated space heater, and in my M.Sc. thesis research I investigated the influence of loudspeaker placement on human music perception.	Rather than flattering the target, present the aspects of your background that make you suitable for the position.
I am very impressed by your research on the dynamics of ketone photoreactions, and would like to be a member of your team as a post-doc.	During my Ph.D. research, I acquired experience in conducting experiments with fs lasers, which should be helpful in studying ultrafast photoreactions in your lab.	Flattery by a job applicant does not sell the candidate. Describing relevant skills and experience does.

9.4.6 Interviewing well

Most industrial research job offers and tenure track academic appointments will only be made after an all-day on-site visit and interview. Many post-doc positions may involve a tele-interview. Job candidates with a Ph.D. degree generally have considerable documentation attesting to their research ability, i.e. their thesis, published papers, recommendations, etc. The interview is used to see if the individual matches his or her paper image, and to form an impression of the candidate as a person. The latter includes ability to communicate verbally and general ability to fit into the organizational atmosphere.

Tele-interviews for postdoctoral positions are typically short, e.g. 10–60 minutes. There is no fixed format, and the candidate needs to follow the lead of the interviewer. You should be prepared to talk about your research to date and your research and career ambitions, and to ask questions about the research directions of the interviewer's group, as well as practical matters (amount of fellowship, cost of living in the locality, etc.) Some interviewers are adept in leading the conversation, and some are awkward. In the latter case, you should initiate describing yourself and your work, as well as asking your questions.

Onsite visits for an academic appointment will usually include giving a departmental seminar, and meeting individually with interested faculty members. Visits in industry, particularly in industrial research departments, will also often include a seminar. Needless to say, the seminar should be well prepared and well executed, following the guidelines given in Chapter 4 of this text.

Seminar and interview performance are critical for competitive positions. Both in industry and academia, the ability to communicate strongly and to interact well with others is important, particularly in career paths leading to leadership positions. Thorough rehearsal of seminars, and practice sessions for post-seminar question and answer periods and interviews, may help a candidate significantly.

University hiring decisions are typically multilevel, and often involve a collective consensus of the senior members of an academic department or its search committee. Accordingly, it is advisable to be familiar with the department and its members, so that you can discuss their research.

During an industrial site visit, the candidate will generally be interviewed by someone in the personnel or human resources (HR) department, the prospective direct manager, and usually also one or more higher level managers. Often, these people will have quite different agendas. Typically, the HR person is interested in the candidate meeting some threshold conditions, and that overall the company's hiring is in line with their diversity and non-discrimination policies. The HR person is the right address for questions about employee benefits. The direct manager is often interested in hiring someone who will get the immediate job done, so that his or her department and the manager personally will look good in the eyes of the organization. While being interviewed by a prospective direct manager, you need to convince him or her that you are the person who will get that job done. The higher level managers, on the other hand, often have a wider view, and in fact are looking to hire the person who will eventually replace that direct manager. The senior manager will be looking for a candidate who can look further towards the future and understands the broader aspects of the company's activities.

In both industrial and academic settings, the onsite visit will often include a tour of the facilities, and probably lunch. If the candidate has any dietary restrictions, it is preferable to notify the hosts about them in advance. Lunch and tours, as well as coffee and mingling periods before a seminar, are opportunities for the organization to know the candidate as a person. Speaking directly with all present and asking good questions about what is seen during a tour, gives a good impression and suggests that the candidate might get along well with his future co-workers.

Both the formal and informal interactions with a potential employer can strongly influence the job decision. As you prepare for a job search you will probably find it helpful to read further on this subject.

For further reading:

Peter J. Feibelman, *A PhD Is Not Enough, A Guide to Survival in Science*, Revised Edition, Basic Books, Perseus, 2011.

Karen Kelsky, *The Professor Is In: The Essential Guide To Turning Your Ph.D. Into A Job*, Three Rivers Press, 2015.

Julia Miller Vick and Jennifer S. Furlong, *The Academic Job Search Handbook*, Fourth Edition, University of Pennsylvania Press, 2008.

Chapter 10

Writing Well: Organization, Grammar and Style

10.1 Writing Discipline

10.1.1 *Begin to write and then keep going*

The old adage, practice makes perfect, is as true for writing as it is for any other skill. Your first research paper may take months to write and to rewrite, again and again. The more you write, the easier and more enjoyable it will be. But first, you must get started, and then, you must keep going. Both can be a challenge. Here are some suggestions:

- Schedule a regular time to write. Identify a time of day when it is easiest for you to concentrate. Some researchers find momentum is maintained best with a routine of writing almost every day, even if it is only for an hour. Others find that a longer block of time is more productive, and they prefer to write for several hours a few days a week. A compromise is to schedule two to three hours of writing time, three to four times a week.
- Treat your scheduled writing time as seriously as any other obligation. Yes, you can, and usually should, say that you are busy, if invited to a meeting or social event at a time you have allotted to writing. Of course, if your thesis advisor can only meet you this week during that time, or it is your daughter's birthday party at nursery school, you probably will decide to change your writing schedule. Note: not cancel, reschedule — preferably on the same day, certainly in the same week.
- Find a place where you have minimum distractions from other people and from other tasks. Working at home first thing in the morning is a traditional pattern amongst writers, but may not be practical for the parent of young children. A quiet area of the university library can be ideal. An office or shared department workspace tends to work best

for those who are able to work at unconventional times when there are few other users around.
- During your writing time, avoid the telephone, e-mail and social networking apps like the plague.
- Set writing goals and monitor your progress. Depending on career stage, goals may include major term papers for coursework or seminars, sections of a thesis, a journal paper, reviewing papers written by others, or even writing a book. The more quantitative the goal, the easier it is to monitor and the better it will help you stay on track.
- Give yourself a reward when you meet a goal. Since you will hopefully be meeting many goals, it is best to pick a variety of modest rewards such as a cup of coffee, a walk in the park with a friend, or a new accessory for your work space. Just don't skip your writing time as a reward for achievement.
- Work on outlines for your main projects, and keep reworking them as you go along. While formulating an outline, you can think about what you are presenting, and according to what logic, without getting bogged down with the exact wording.
- Start with the sections of the research paper that are easiest to write. Most beginning researchers find describing apparatus and procedures relatively straightforward. Others find that starting to write the literature review while conducting the literature search is helpful. For many, writing the results section first delineates what is needed in all of the other sections.
- Consider stopping your writing session in the middle of a section if continuing with familiar material helps you get started more quickly at your next writing session.
- Write quickly, even if sloppily, to facilitate the flow of ideas from the brain to the computer. Later, the text can be reworked to improve readability.
- Consider forming a mutual support group with colleagues who are facing the same writing challenges. These are similar in concept to groups that focus on other common interests such as sports training, losing weight or parenting. The most successful groups focus on sharing concrete goals and cheering each other on to meet them. The

group can also be a forum for sharing resources and learning about good writing.

10.1.2 *Organizing materials*

10.1.2.1 *Document naming and tracking*

A system for saving and keeping track of relevant documents related to each research project is essential. These documents may include laboratory notes, records of discussions with colleagues and advisors, drafts of research papers and the final edited version of the paper. Electronic filing takes no physical space and it is easy to adapt as your needs change. However, it can become even more overwhelming than a roomful of papers if not well organized and maintained. The best filing system is one that makes sense to you and your collaborators. It must be easy to maintain over time. Define folders, sub-folders and document names that allow quick identification of a needed document or e-mail during research, while writing the research report, and even after publication. Choose titles related to the content of the project that will be understandable to all of your collaborators. Although most word processors keep track of revision dates and versions, consider incorporating either the version number or the date, as well as the initials of whoever made the last revision, into the file name.

10.1.2.2 *Backups*

It is all too common to postpone backing up critical documents. Every disk drive eventually fails and inevitably does so at an inconvenient time. Working routinely in a cloud storage environment is a good but not necessarily sufficient solution. The more cautious approach is to update an additional set or sets of backups of critical documents on external devices under your sole control. While writing, use the automatic save feature of the word processor to automatically save your work at least every hour (every 15 minutes is better), and back up your work at least daily onto an independent device (e.g. a memory stick and/or an external drive).

10.1.2.3 *Versions and collaborators*

The near universality of electronic communication with collaborators and the ease of updating documents often means that each paper will go through many versions. Take special care that each of the co-authors works on the same version at any given time. Various cloud storage facilities provide an easy way to share files among those involved on any given project. However, it is still best to maintain backups of at least the latest version on your own computer and on an external device.

10.1.2.4 *Hard copy storage*

Some writers still prefer to review a printed paper document, a few even prefer handwriting their research reports. If you are among them, find a secure way of saving and labeling your work. It is good practice to include the title, date, and page in the header or footer of each page.

10.1.3 Word-processor tools

Thesis writing, in particular, is very time consuming. Good writing usually demands a great deal of rewriting, especially for the beginning academic scientific writer. A few easily learned tools minimize the dirty work and give you more time to think about content.

If you didn't learn to touch type when you were a secondary school student, it is not too late. There are many free online teaching applications. A weekend or several evenings spent on the basics will enable you to start touch typing in your day to day writing, and with daily practice, your speed will increase. The time saved will more than repay the time invested in learning to touch type.

Several advanced word processing features can save considerable time in scientific and technical writing tasks. Word processors vary in specific features and how they are implemented. Furthermore, the details of their implementation change with successive versions of word processing programs. The function of the most useful of these features, using the nomenclature of Microsoft Word, is noted in Table 10.1. Use the help function to learn how to implement each feature on your word processor.

Table 10.1. Word processor features useful in research reports. The location of the command for each in Microsoft Word 2016 is indicated in parenthesis under the feature name. Use the help feature to learn details and to learn how to implement in other word processors.

FEATURE	FUNCTION	COMMENTS
Tab (home/ paragraph)	Starts the following text at a preset horizontal position.	Advanced tab features can control horizontal position with respect to right margin, center, and left margin, and can align according to the decimal point in numbers. Do not insert blank spaces to control horizontal position when using a word processor.
Indent (home/ paragraph)	Sets margins of a paragraph for something other than the default.	Use to emphasize text. e.g. for direct quotations.
Space before, space after (home/ paragraph)	Leaves a pre-determined vertical separation between paragraphs.	Do not use insert blank lines by pressing the ENTER key to control vertical spacing when using a word processor.
Widow-orphan control (home/ paragraph)	Prevents a single line of a paragraph from appearing on one page (widow), and the rest (orphan) on the succeeding pages.	Use with body text, as well as headings.
Keep together (home/ paragraph)	Keeps elements of a paragraph (any unit of text which ends with touching the return key, including headings and regular text paragraphs) from being split between two pages.	Use with headings to prevent part of the paragraph appearing on the bottom of one page, and part on the top of the next page.
Keep with next (home/ paragraph)	Keeps a paragraph (any unit of text that ends with touching the return key, including headings and regular text paragraphs) together on the same page with the next paragraph.	Use with headings to prevent the heading from appearing on the bottom of the page, without the text paragraphs which it heads. Note that this feature will not work if blank lines were inserted by touching the return key.

FEATURE	FUNCTION	COMMENTS
Sticky-space (Ctrl-Shift-Space)	Inserts a space that will not break at the end of a line.	Use between a number and its units (e.g. *5 km/hr*), between initials and a last name (*A.B. Yehoshua*), and within numbered items (e.g. *Table 2*) to prevent the first part being at the end of one line, and the second part being on the beginning of the next line.
Sticky-hyphen (Ctrl-Shift-Hyphen)	Inserts a hyphen (or minus sign) which will not break at the end of a line.	Use for in-line mathematical expressions to prevent part of the expression being at the end of one line, and the other part at the beginning of the next line.
Table (insert/table) (layout)	Provides vertical and horizontal arrangement of any combination of text and graphics.	Use to arrange a matrix of photographs, figures, etc., together with their explanations or captions. Format the table to prevent rows from breaking across pages, and to repeat the header row if the table is on multiple pages.
Automatic numbering of references, figures, tables, and equations (references)	Automatically assigns and reassigns reference, figure, table and equation numbers.	Allows inserting, removing, or re-arranging references etc., without manually renumbering.
Cross reference (references/ cross reference)	Allows inserting, removing, or rearranging figures, equations, etc., without individually correcting the references to them in the body text.	Use when citing an existing reference, or in the text in referencing a figure, table, or equation.
Manage sources (references/ manage sources)	Manages citations.	Use to search for sources in previous papers and to format the bibliography.

FEATURE	FUNCTION	COMMENTS
Outline view (view)	Allows preparing, viewing and manipulating the organization of a composition.	Prepare and edit the outline using outline view. Each element of the outline automatically becomes a heading of the composition at the appropriate level.
Table of contents (references/ toc)	Automatically prepares a table of contents based on headings.	Useful for long compositions such as theses and books.
Style (home/ styles)	A collection of formatting instructions for font and paragraph (e.g., font size, shape, and enhancements, indentation, space before and after, etc.) to be used for a specific type of text.	Use predefined or user-defined styles for body text, each level of heading, captions, equations and references.
Template (file/new to open existing template)	A collection of defined styles and standard text ("boiler plate").	Prepare a template for each type of frequently encountered writing genre, to avoid duplicating the work of defining styles.
Track changes (review/ track changes)	Allows collaborators and supervisors to reversibly edit and revise.	Revisions can later be accepted or rejected. Allows efficient collaboration.
Comments (review/ comments)	Places comments within balloons in the margin, attached to marked text.	Allows efficiently exchange of opinions and comments related to a specific text, but not part of it.
Spell checker (review/ spelling and grammar))	Checks spelling against a built-in vocabulary list.	Many technical terms are not included in the built-in vocabulary list, and will need to be correctly entered manually. The spell checker will not detect a word as misspelled if it happens to be a different correctly spelled word (e.g. *The device processed one peace at a time. Peace* will not be noted as erroneous, although *piece* was intended.)

FEATURE	FUNCTION	COMMENTS
Grammar checker (review/ spelling and grammar))	Check for compliance with a list of grammar rules.	It is good practice to run the grammar checker and to consider, but not necessarily to implement, each suggestion. Some suggestions are not appropriate for scientific writing. In particular, the suggestion to use active voice should usually be disregarded.

10.2 Composition

In Chapter 2, the conventional sections and sub-sections of the research report were explained. The major challenge in organizing these sections and sub-sections is consistency — keeping to a logical order and avoiding misplaced statements (method statement in Results, results in Discussion, etc.) In this section, the building blocks of all English composition, i.e. the paragraph and the sentence, will be discussed. Then, in section 10.3, basic principles of English grammar relevant to scientific writing will be outlined and guidance on word choice will be given. Presenting content clearly, in a structured fashion, is necessary for all writing to be readable. Structural mistakes and logical deficiencies in presenting material are typically the most common error amongst both native and non-native English speakers.

10.2.1 *Hierarchical structure*

Scientific writing has a top-down hierarchal structure as illustrated in Figure 10.1.

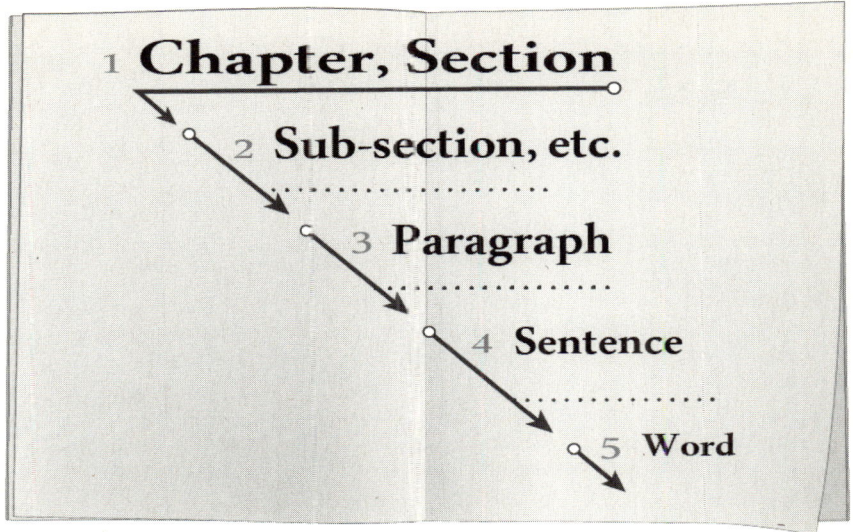

Figure 10.1. Hierarchical order in English composition.

10.2.2 *The paragraph*

A good paragraph systematically develops a topic. The first sentence defines the topic of the paragraph. The subsequent sentences develop the topic in logical order. The final sentence presents the conclusion or the most important point. Repeating key words and referring back to old information improves paragraph flow and helps the reader follow the development of the topic. In essence, the paragraph is a mini-composition.

Typographically, the start of the paragraph is announced to the reader either by indentation (best) or by a double space between lines. This visual signal alerts the reader to the introduction of a new topic. The paragraph should have at least two sentences, more are preferred. On the other hand, paragraphs that are too long can confuse a reader. Three to seven sentences in a paragraph is a good general rule.

Many books on scientific writing recommend preparing and working according to an outline. Some writers view an outline as good in theory but they find that they rarely use one. Considering how widespread structural and logical errors are among beginning researchers, we highly recommend preparing a detailed outline that goes down to the level of defining the topic of each paragraph. It may be helpful to note, in bullet point form, the steps in the development of the main idea of the more complicated paragraphs. Examples of paragraph construction are given in Table 10.2.

Table 10.2 Paragraph examples

Paragraph example	Explanation
Both tests show that the etalon window provided better microwave leakage attenuation than the original grid window, as well as vastly better visibility. Although the metal grid window provides sufficient attenuation to meet current safety standards, we may anticipate efforts to lower the maximum allowable leakage due to interference with communications using the 2.45 GHz ISM frequency (e.g. WiFi and Bluetooth). Most importantly, the etalon window provides cook and look feedback to assist the consumer in correctly preparing food.	←The first sentence introduces paragraph topic (etalon window) ←The next sentence develops the topic (characteristics of the window) ←The concluding sentence conveys the main point (most important characteristic of the etalon window):
The distribution of the spots between the roof and slope surfaces was calculated as follows. Each HVSV recording contained 17 117 frames, and each frame showed the cathode spot locations on all the cathode surfaces. The images from all these frames were then integrated. The integrated distribution was generated by purpose-designed "Spot view" software, written by Shoval[9] in MATLAB. "Spot view" divided the image into square bins having a length of 3 pixels where each pixel had a physical size of 0.33 mm in the x and y-directions, so that each square bin represented a projected square with a surface area of 0.1089 mm².	←The first sentence introduces the paragraph topic – calculation of spot distribution. ←The following 3 sentences describe the calculation process. ←The concluding sentence conveys the main point.

Paragraph example	Explanation
The Er_2O_3 coatings were annealed in vacuum (~1.33 Pa) at 1273 K for 1 hour. The annealing changed the crystallographic structure of all coatings deposited on RT substrates from nearly amorphous (Fig. 3) to crystalline $C-Er_2O_3$ (Fig. 7). The latter diffraction data is typical of all annealed coatings on RT substrates. No β-Er_2O_3 peaks were found in the post-annealing XRD patterns. Texture analysis (Table 1) showed that the annealed coatings did not have a preferred orientation, although the coatings had dominant {211}, {111} and {332} crystallographic planes. For coatings deposited on 673 K heated substrates, the annealing increased the peak intensity, while keeping and increasing their {111} preferred orientation (Table 1). The coating grain size, after annealing, was 450 ± 50 nm and 380 ± 50 nm for coatings deposited at RT and T=673 K, respectively.	←The first sentence introduces the paragraph topic: effect of annealing on Erbia coatings. ←The next 4 sentences report on the results. ←The concluding sentence summarizes the most important finding.

10.2.3 The Sentence

The sentence expresses a complete thought. It is the basic building block in the hierarchical organization of the paper. Ideally, each sentence will contribute logically and informatively to the development of the main point of the paragraph.

A sentence must have a subject and a predicate.

- The subject is the noun (or noun phrase or pronoun) that performs the verb (the do-er or the be-er) in active sentences, and the receiver of the action in passive sentences. A <u>noun phrase</u> comprises a "head noun" with one or more modifiers (see section 10.3.2), e.g. <u>A fast transport mechanism</u> [subject] *transferred the samples into* <u>the process chamber</u>.
- The predicate consists of the verb and any accompanying modifiers along with other words that receive the action of the verb or complete its meaning. The verb indicates the action performed (e.g. *accelerate*); a feeling (e.g. *doubt*); or a state of being (e.g. *is*). The predicate may also include:

- Words or phrases that complete the meaning of the sentence by describing or renaming the subject. These usually follow verbs that express a state of being (stative verbs). [e.g. *The vacuum chamber was <u>rusty</u>; The tensile strength of the weld was <u>highest when the laser was scanned at 1.3 cm/s</u>.*]
- Additional words or phrases that complement action/feeling verbs. These may:
  - Describe something about the subject [*The vacuum chamber was constructed <u>from 304 stainless steel</u>.*]
  - Tell us who or what receives the action (direct object) [e.g. *A diffusion pump evacuated <u>the chamber</u>.*]
  - Indicate to or for whom or what the action is performed (indirect object). In scientific writing, a prepositional phrase is more apt to fulfill the function of an indirect object. [e.g. *The diffraction pattern was recorded <u>for a zinc oxide thin film</u>.*]
  - Explain how the action was accomplished [e.g. *The electron density was measured <u>interferometrically</u>.*]

Sentences may have more than one subject and more than one predicate joined by connecting words. For example: *The data <u>were collected</u> and <u>(were) analyzed</u>. <u>The treated samples</u> and <u>the control samples were analyzed</u> and <u>(were) compared</u>.* A complex sentence structure can connect two or more related points more succinctly than two or more simple sentences. However, complex sentences are often harder for the reader to understand. Moreover, the more complex the sentence, the more difficult it is for the novice writer or the non-native English speaker to compose it correctly. Some guidelines for building good sentences, as well as examples of common errors and fixes, are described below.

10.2.3.1 *Use the natural English word order*

Follow the natural word pattern unless there is a good reason to do otherwise. Typical word order varies from language to language. The natural order in English sentences consists of the subject noun and its modifiers, the verb and its modifiers, and everything else. This is formally termed Subject-Verb-Object. The verb normally follows the subject. Everything else may include direct objects, indirect objects, prepositional phrases, predicate adjectives, predicate nominatives, and complements. In scientific writing, most of the sentences should be structured in natural word order.

Some languages have a different natural word order. For example, Subject-Object-Verb order is used in Armenian, Hindi, Hungarian, Korean, Japanese, Persian and Turkish and sometimes in Chinese, Dutch and German. If your native language is one that is based on Subject-Object-Verb order, it is particularly important to take care to build sentences according to the natural English word order. Some examples are given in Table 10.3.

Table 10.3. Sentence examples — preference for natural word order.

No!	Yes!	Explanation
When x>r, this relationship is valid.	This relationship is valid when x>r.	Focuses on the subject
Using Langmuir and emissive probes, radial profiles of the plasma density, electron temperature and plasma potential were measured, for various gas flow rates.	Radial profiles of the plasma density, electron temperature and plasma potential were measured at various gas flow rates with Langmuir and emissive probes.	Emphasizes what was done, rather than how it was done.
During heart transplantation using cardiopulmonary bypass heart-lung machines, patients exhibit an inflammatory response that is characterized by increased expression of at least ten leukocyte cluster-of-differentiation antigens.	Patients exhibit an inflammatory response that is characterized by increased expression of at least ten leukocyte cluster-of-differentiation antigens during heart transplantation using cardiopulmonary bypass heart-lung machines.	Emphasizes what happened.
Compared with the other major light sources, the incandescent bulb is the least efficient.	The incandescent bulb is the least efficient major light source.	Emphasizes the subject.

A good reason in scientific writing to use a different word order is to put old information before new information. In the first example in

Table 10.3, the "no" sentence [*When x>r, this relationship is valid.*] could be more appropriate if the relationship between x and r were explained previously (i.e. is old information).

10.2.3.2 Construct complete sentences

Avoid incomplete constructions or fragments. A sentence must express a complete thought and must have a subject/verb pair. (Imperative sentences, such as the first sentence of this paragraph, have an implied subject "you". However, imperative sentences are rarely used in scientific writing.) Sentence fragments are common in text messages and even in e-mail. They are used in lecture slides, posters, and captions. However, they have no place in the body text of formal academic writing. Some examples of incomplete and complete sentences are given in Table 10.4.

Table 10.4. Sentence examples – incomplete and complete constructions.

No!	Yes!	Explanation
Which can reduce micro-grid's dependence of reserve capacity from the external grid.	The optimization power distribution method can reduce the dependence of the micro-grid on reserve capacity.	The subject is missing in the no example.
Now estimated the particle flux Γ_{Nf} and average velocity v_{Nf} of this fast component of the neutral flow.	The particle flux Γ_{Nf} and the average velocity v_{Nf} of the fast component of the neutral flow were estimated.	The subject is missing in the no example. The yes example also uses natural word order.
The unsuccessful may be attributed to difficulties in attaining strong receptor-ligand binding.	The unsuccessful trials may have been caused by failure to obtain strong receptor-ligand binding.	The subject is missing in the no example.
A mode of operation which is responsible for TE_{01} excitation called a normal mode.	The normal mode of a helical antenna excites the TE_{01} mode in circular waveguides.	The main clause of the 'no' example is missing a verb.

10.2.3.3 Use natural verb forms

Speakers of English as a second language, in particular, tend to over-use nouns derived from a verb plus a generalized verb. Use the natural verb form to simplify the sentence and make it more precise and powerful. For example, in the sentence "<u>Data collection</u> <u>was accomplished</u> with a digital-to-analog converter and a personal computer.", collection is a noun form of the verb collect and *was accomplished* is a generalized verb. It would be better to say, <u>Data were collected</u> with a digital-to-analog converter and a personal computer. This sentence is both shorter and stronger than the first version. See Table 10.5 for more examples and section 10.3.1.4 for further guidance on commonly used verbs in scientific writing.

Table 10.5. Sentence examples — use of natural verb form.

No!	Yes!	Explanation
Measurements were made of the coating hardness using a nano-indenter.	*The coating hardness was measured using a nano-indenter.*	Use the natural verb form *measure* rather than *Measurements were made*
In this paper, a numerical investigation was performed of the dielectric breakdown properties of hot SF$_6$ gas contaminated by copper at temperatures of 300–3500 K and pressures of 0.01–1.6 MPa.	*In this paper, the dielectric breakdown properties of hot SF$_6$ gas contaminated by copper are quantified at temperatures of 300–3500 K and pressures of 0.01–1.6 MPa.*	Use the natural verb form *quantified* rather than *a numerical investigation was performed.*
An analysis was performed on the data.	*The data were analyzed.*	Sentence reduced from 7 to 4 words, by using analyzed as a verb.
A high-speed video camera was used to make observation of the explosive dynamics.	*A high-speed camera was used to observe the explosive dynamics.* or *A high-speed camera recorded the explosive dynamics.*	The natural verb form (*observe*, *record*) states what was done.
Evacuation of the chamber was performed by a diffusion pump.	*The chamber was evacuated with a diffusion pump.* or *A diffusion pump evacuated the chamber.*	The choice of subject [chamber or diffusion pump] depends on the desired emphasis.

10.2.3.4 Focus on the main thought

Choose the subject and verb to emphasize the main thought. Put the main thought in an independent clause. Put secondary information into subordinate clauses or phrases. An independent clause expresses a complete thought and can stand alone as a sentence. A subordinate clause or phrase does not express a complete thought and cannot stand alone as a sentence. Examples are given in Table 10.6.

Table 10.6. Sentence examples — subordinate clauses.

No!	Yes!	Explanation
The control capacity of DFIG was fully utilized, which resulted in improved system damping without degradation in the quality of the voltage control.	*System damping can be improved significantly without degradation in the quality of voltage control by fully using the control capacity of DFIG.*	Focuses on the improvement due to the new method, rather than the comparison to the existing method.
A phase-locked loop was used to control the mirror position, which was critical to achieving stable operation.	*Stable operation was achieved by controlling the mirror position with a phase locked loop.*	Focuses on achieving stable operation (the goal), not the method.

Avoid beginning a sentence with a long introductory phrase. This structure tends to put the cart before the horse — i.e. the reader has to get through the secondary detail before reaching the main point of the sentence. Examples are given in Table 10.7.

Table 10.7. Sentence examples – avoid long introductory phrases.

No!	Yes!	Explanation
Using a CSEM model 3400 nano-indenter equipped with a flashlight and a microcomputer, the hardness of the coating was measured.	Coating hardness was measured using a CSEM model 3400 nano-indenter equipped with a flashlight and a microcomputer.	The main point is that the hardness of the coating was measured, the means are secondary.
For Ar gas flow rates of 13–100 sccm, discharge currents of 1–1.9 A, and magnetic field intensities up to 160 G, radial profiles of the plasma potential, the electron temperature and the plasma density were measured.	Radial profiles of the plasma potential, the electron temperature and the plasma density were measured for Ar gas flow rates of 13–100 sccm, discharge current of 1.0–1.9 A, and magnetic field intensities 0–160 G.	The main point is what was measured.

10.2.3.5 *Use only necessary words*

During review, analyze each sentence to eliminate every word that is not needed to convey the intended meaning. Several common categories of unnecessary words are:
- Wordy phrases that mean little e.g. *as a consequence of, by means of, despite the fact that, for the purpose of, in the final analysis, in the event that, in the process of, in order to.*
- Unnecessary qualifiers. Eliminate adjectives and adverbs that rank a result. Let the reader conclude what is *very* important or *remarkable*.
- Redundancy. Don't say the same thing twice. For example, *process* is a much-overused word that should not be combined with another word that itself defines a process. e.g. *deposition*. Thus *deposition process* is redundant, *deposition* is sufficient.
- Complex constructions. Sentences can usually be shortened by eliminating the phrase *there is (are)*. Long prepositional phrases can often be eliminated by incorporating the main information within the subject.

Examples of eliminating unnecessary words are presented in Table 10.8.

Table 10.8. Sentence examples — avoiding unnecessary words.

No!	Yes!	Explanation
Wordy phrases		
In addition to others, chromium nitride is one of the popular and universally used coating systems.	Chromium nitride is among the frequently used coating systems.	The phrase *in addition to others* is vague and provides no information. If others are significant, they should be mentioned specifically.
The images that you see in the video clip of the biomechanics of breastfeeding are familiar.	Familiar images are presented in the video clip of the biomechanics of breastfeeding.	The phrase *that you see* is unnecessary.
The voltage was measured by means of an oscilloscope.	The voltage was measured with an oscilloscope.	The phrase *by means of* does not convey any additional meaning.
Specifying the cathode properties in this particular case will give a better understanding of the material properties critical to determining the re-ignition of the spots, which is crucial to our understanding of the maintenance of the spots.	This research is expected to help us better understand the requirements for maintaining cathode spots.	The sentence is simplified by focusing on the ultimate goal of the research and deleting a few bombastic phrases: *in this particular case, critical to determining, which is crucial.*
The value of the voltage was set to 5.1 V.	The voltage was set to 5.1 V.	No additional meaning is conveyed by *the value of.*
Microwaves can propagate through free space allowing remote sensing to be accomplished.	Microwaves can propagate through free space for remote sensing.	No additional meaning is conveyed by *to be accomplished.*
Solid dielectric materials are opaque to light and infrared radiation but transparent to microwaves, which permits the probing of the whole volume of objects transported inside a dielectric tube without the need for inserting windows.	Solid dielectric materials are opaque to light and infrared radiation but transparent to microwaves, which permits probing the whole volume of objects transported inside a dielectric tube without windows.	No additional meaning is conveyed by *the need for.*

No!	Yes!	Explanation
It should be obvious from Figure 5 that the voltage saturated at 5.7 V after 23 μs.	Figure 5 shows that the voltage saturated at 5.7 V after 23 μs.	*It should be obvious* and similar expressions like *as we can clearly see, it was clearly demonstrated* are unnecessary when it is obvious and annoying to the reader if it is not so obvious.
	Unnecessary qualifiers	
The sources are of particularly small dimensions if a filtering of macro-particles is not required.	The sources are particularly small if macro-particle filtering is not required.	The words *of* and *dimensions* are unnecessary. Using the noun phrase *macro-particle filtering* is clearer than *a filtering of macro-particles*.
The hard evidence presented here validates Miller's theory.	The results in Figs. 3–5 validate Miller's theory.	Be specific and let the reader judge if the evidence is hard.
The experiments presented here are truly substantial, and led to really important conclusions.	The measurement of relativistic electron energy loss provide design rules for a high power THz free electron laser.	Be specific and let the reader judge what is substantial or important.
	Redundancy	
The desired properties of chromium nitride coatings mainly depend on their microstructure, which can be directly influenced by varying the level of the deposition energy during the deposition process.	The properties of chromium nitride coatings depend on their microstructure, which depends on the energy of the depositing particles.	Redundant phrase: *during the deposition process*. The clause *which can be directly influenced by varying the level of the deposition energy* is expressed more succinctly in the Yes example.

No!	Yes!	Explanation
The problem of computing fCM from sCM was previously addressed by Honey at. el. in [1], however, the model failed to address the important step of normalizing sCM.	*Honey et al [1] computed fCM from sCM, however their model did not normalize sCM.* *If it is not important for the citation to be author prominent: Although fCM was computed from sCM, the model did not normalize sCM. [1].*	Duplication is eliminated and the passage shortened by using *computed* rather than *the problem of computing*, deleting *was previously addressed* (implicit), and stating that the model *did not normalize* rather than it *failed to address the important step of normalizing*.
The designed control strategy provides an energy-efficient way for the continuous sterilization, and might reduce the cost for the plants that are involved in the continuous sterilization process.	*The control strategy presented in this research improves the energy efficiency of continuous sterilization and might decrease its cost.*	The sentence is simplified by eliminating duplicative words and phrases that unnecessarily complicate the text and do not contribute to the main points.
Complex constructions		
In an electromagnetic wave, there is an electric field that can push or pull electrical charges.	*The electric field in electromagnetic waves can push or pull electrical charges.*	Deleting *there is* and rewording simplifies the sentence.
Producing accurate sensors for personal levels of ultraviolet radiation on zinc oxide would provide affordable electronic devices for monitoring and achieving a balance of the health benefits of synthesizing sufficient levels of vitamin D while reducing the risk of developing melanoma skin cancers.	*Accurate zinc oxide ultraviolet radiation sensors would enable affordable personal monitoring, for balancing the health benefits of vitamin D synthesis against the risk of melanoma.*	Superfluous words are deleted by using a detailed noun phrase *Accurate zinc oxide ultraviolet radiation sensors* (the device) as the subject and focusing the predicate on what benefit the device may provide.

No!	Yes!	Explanation
What are the differences of the dielectric properties between pure epoxy and epoxy/crepe composites?	How do the dielectric properties of pure epoxy and epoxy/crepe composites differ?	Wordiness is reduced by making *dialectic properties* the subject and using the specific verb *differ* instead of *What are the differences....?*
The using of the virtual reserve capacity trading optimal power distribution method for the power system containing micro-grids, combined with the existing optimization objective function of minimum of operation fuel consumption and minimum the total amount of daily purchased power is scientific and reasonable.	A power distribution optimization method for virtual reserve capacity trading in power systems containing micro-grids was developed which maximizes use of storage devices to minimize fuel consumption and the power purchased daily.	Sentence simplified by focusing on what was done.

10.2.3.6 *Be precise*

Describe apparatus, methods and results accurately and, where appropriate, quantitatively. In Methodology and Results, it is especially important to be specific so that the research can be replicated. For example, state how many items were used rather than indicating *several* and note time intervals rather than writing *frequently* or *often*. Some examples are given in Table 10.9.

Table 10.9. Sentence examples — using precise terms.

Imprecise	More precise	Explanation
The anode flanges had several holes for attaching screws that held it together with the base plate and the chamber extension.	The anode flanges had 16 bolt-holes to attach it to the base plate and the chamber extension.	*Several* quantified. *Screw* is probably inaccurate – more likely *nuts and bolts* were used.

Imprecise	More precise	Explanation
The current waveform had a very fast rise time.	The current rose from zero to 90% of the steady-state value of 11.3 A in 23 µs.	Current and time quantified.
The annealed sample had very large grains.	The grain size in the annealed sample was 1.3 µm.	Grain size quantified.

10.2.3.7 Compact the text by using compound sentences

Combine two sentences with shared components to save words and improve paragraph flow. Substituting a compound sentence is especially appropriate when the original two sentences are short and closely related, as illustrated in Table 10.10.

Table 10.10. Sentence examples — compound sentences.

Long Form	Shortened Form	Explanation
The data were collected. Then the data were analyzed.	The data were collected and analyzed.	The subject [*The data*] is common in both of the long form sentences. The second [*were*] is assumed in the shortened form.
The data were collected and correlations were calculated.	The data were collected and correlations calculated.	The second *were* is assumed.
The samples which were produced were examined microscopically.	The samples produced were examined microscopically.	*Which were* is assumed.

10.2.3.8 Use similar constructions to express similar situations

Use parallel structure to emphasize similarity in content. This means using similar grammatical forms for similar phrases within a sentence or group of sentences, or in a list. The first example in Table 10.11 below illustrates this principle. The second example shows an extension of this principle.

Table 10.11. Sentence examples — similar constructions.

No!	Yes!	Explanation
Thin film properties can be improved by incorporation of alloying elements, annealing, and the use of accelerated ions.	Thin film properties can be improved by incorporating alloying elements, annealing, and accelerating incident ions.	All elements in a list should have the same grammatical form.
The samples were prepared by the following procedure. First, each element was washed in alcohol. Then a high speed air jet was used to dry the elements. Finally, mounting the elements was accomplished using an epoxy adhesive.	The samples were prepared by the following procedure. First, each element was washed in alcohol. Then, the elements were dried in a high speed air jet. Finally, they were mounted using an epoxy adhesive.	Writing each step in the procedure in the same form emphasizes that each step is part of the same procedure.

10.2.3.9 Use consistent terminology

Use straightforward and consistent terms in scientific writing to make the reader's job easier. The examples in Table 10.12 below demonstrate this principle.

10.2.3.10 Use sentence structure to improve paragraph flow

Vary short and long sentences within the paragraph. Shorter sentences often work well to introduce the main topic and to draw the reader into its development. Longer sentences may be necessary to express complex ideas. Following a complex sentence with one or more short and simple ones helps the reader.

Table 10.12. Sentence examples — consistent terminology.

No!	Yes!	Explanation
The pressure flange was held together by eight 10 mm diameter non-magnetic stainless steel <u>bolts</u>. Each <u>screw</u> was first tightened by manually rotating a nut until finger-tight, and then using a wrench to further rotate them one full turn. In this way, the <u>fasteners</u> were stretched by approximately 0.004%.	The pressure flange was held together by eight 10 mm diameter non-magnetic stainless steel <u>bolts</u>. Each <u>bolt</u> was first tightened by manually rotating a nut until finger-tight, and then using a wrench to further rotate them one full turn. In this way the <u>bolts</u> were stretched by approximately 0.004%.	In the "no" example, the reader may be uncertain if *bolt*, *screw*, and *fastener* refer to the same item. Using *bolt* consistently in the "yes" example and in any accompanying diagrams eliminates possible ambiguity.
Microwaves were generated by a 2.45 GHz <u>signal source</u> based on a magnetron tube. The <u>generator</u> was coupled to the processing chamber by a WR340 waveguide. The <u>oscillator</u> input voltage and load impedance were stabilized to insure frequency stability to within 2.3 MHz.	Microwaves were generated by a 2.45 GHz magnetron used as the <u>signal source</u>. The <u>source</u> was coupled to the processing chamber by a WR340 waveguide. The <u>source</u> input voltage and load impedance were stabilized to insure frequency stability to within 2.3 MHz.	In the "no" example it is not clear if *source*, *generator*, and *oscillator* refer to the same object. Using only one of these terms [*source*] consistently, helps the reader understand that each sentence refers to the same object.

10.2.3.11 *Habitually improve sentences*

Do not underestimate the value of editing your own writing. Read and reread to consider if each sentence clearly and logically contributes to the development of the topic. Ensure that the meaning is clear, and that the sentence is grammatically correct and concise.

> "Vigorous writing is concise. A sentence should contain no unnecessary words and a paragraph no unnecessary sentences, for the same reason that a drawing should have no unnecessary lines and a machine no unnecessary parts. This requires not that the writer make all his sentences short, or that he avoid all detail and treat his subjects only in outline, but that every word tell."
>
> William Strunk Jr. in Elements of Style

10.3 Choosing your Words

The purpose of this section is to provide guidance on choosing the right words to make your point. We will review selected aspects of English grammar that are particularly important in scientific writing. Although English grammar is complex and English vocabulary is rich, using simple structures and straightforward non-technical words is generally preferable in scientific writing and can help balance the necessary use of complex technical terms. Simplicity is kind to the reader, who may be a non-native speaker of English. Moreover, if English is not your first language, using basic structures and vocabulary can help you avoid awkward mistakes.

10.3.1 *Verbs*

A verb is a word or set of words that shows action *(accelerates)*, feeling *(doubts)*, or state of being *(are, have, seem)*. Properly chosen, the action verb can be the strongest part of the sentence. It clearly expresses exactly what the subject did [*The heavy ions <u>impacted</u> the substrate.*] (in active voice) or what was done to the subject [*The electrons <u>were heated</u> by the electric field.*] (passive voice). State of being verbs are also called stative or linking verbs. They include all forms of the infinitive *to be (am, is, were, are, was)*, plus such words as *seem, look, feel, appear* and *act*. Unlike action verbs, they must link the subject with the predicate modifier, often an adjective but sometimes a noun or pronoun that restates or defines the subject. *Examples:* [Adjective: *The plasma was <u>hot</u>.* Noun: *The chamber material was <u>stainless steel</u>.*]

English verbs often consist of more than one word. For instance, in the sentence [*The increased coating hardness <u>may have been</u> the result of increased nitrogen incorporation.*], the verb is *may have been*. Some of the words in a multi-word verb are called helping verbs, so named because they help clarify the intended meaning. Many verbs can function as helping verbs: *shall, must, do, has, have, can, may, should,* and *could*.

The form of English verbs depends on tense, voice, and number.
- Tense indicates when the action or condition expressed by the verb occurs.

- Voice, active or passive, denotes the relationship between the action or state of being expressed by the verb and the subject or object participants. When the subject is the agent or doer of the action, the verb is in the active voice. When the subject is the receiver of the action, the verb is in the passive voice.
- Number is either singular or plural.

Verb forms and the scientific conventions applicable to them will be explained in more detail in sections 10.3.1.1–10.3.1.3 below.

10.3.1.1 *Verb tenses*

In general writing, tense expresses when the action of the verb occurred, however, in scientific writing it may also convey the authors' attitude towards the generality or specificity of the event. The choice of verb tense is highly conventional in scientific writing. The past tense not only denotes an action that already occurred, but also designates a result that is specific to the conditions under which it was obtained, and not generally applicable. The present tense not only denotes an action that is occurring now, but also designates a result that is generally applicable. The simple past and the simple present are the tenses most used in research reports. The future tense is used mainly in the Discussion section of the research report and the Methodology and Work Plan sections of research proposals. English has many tenses. Table 10.3 summarizes the most common English verb tenses, illustrates their use with a common regular verb, *investigate*, and highlights the tenses used most often in scientific writing.

Table 10.13. English tenses, illustrated in the active voice. Tenses commonly used in scientific writing are in **bold**.

Tense	Example	Meaning
Past	***Smith investigated the influence of pH on the survival of E. coli bacteria.***	**Smith completed this particular investigation. It was reported in one or a few papers. (This is the standard tense in the Literature Review, Methodology, and Results sections).**

Tense	Example	Meaning
Past perfect	Smith *had investigated* the influence of pH on the survival of *E. coli* bacteria.	Smith completed this particular investigation before something else happened. (Perhaps another significant study in the field).
Past progressive	Smith *was investigating* the influence of pH on the survival of *E. coli* bacteria.	Smith was in the process of investigating when someone or something interrupted his work. He may continue the investigation later, but it is not certain if he will do so.
Past perfect progressive	Smith *had been investigating* the influence of pH on the survival of *E. coli* bacteria.	Smith performed the investigation over a period of time and completed it.
Present	**Smith *investigates* the influence of pH on the survival of *E. coli* bacteria.**	**This sentence suggests that this investigation is what Smith generally does.**
Present Perfect	**Smith *has investigated* the influence of pH on the survival of *E. coli* bacteria.**	**Smith performed an investigation that began in the past, continued over a long period, and perhaps is still going on. This investigation probably resulted in a string of publications or a major work. (Use this form in literature reviews to emphasize that the cited works were extensive).**
Present Progressive	Smith *is investigating* the influence of pH on the survival of *E. coli* bacteria.	Smith is performing this investigation at the present moment and is continuing this investigation.
Present Perfect Progressive	Smith *has been investigating* the influence of pH on the survival of *E. coli* bacteria [but he has not yet concluded the study].	Smith has been performing the investigation over a period of time and is likely to be continuing the work in the future.
Future	**Smith *will investigate* the influence of pH on the survival of *E. coli* bacteria.**	**Smith will perform the investigation in the future. (This could be a sentence in the work plan of a research proposal).**
Future perfect	Smith *will have investigated* the influence of pH on the survival of *E. coli* bacteria [before the end of the term].	Smith will complete the investigation by some specific time in the future.

Tense	Example	Meaning
Future progressive	Smith *will be investigating* the influence of pH on the survival of E. coli bacteria [during the spring semester].	Smith will be performing the investigation continuously at some time in the future.
Future perfect progressive	Smith *will have been investigating* the influence of pH on the survival of E. coli bacteria for over four years.	Smith has been performing the investigation over time and will continue to do so. He will probably complete it, but it is not clear when.
Modal auxiliary	Smith *may have investigated* the influence of pH on the survival of E. coli bacteria.	Possibly Smith performed the investigation, and possibly not.

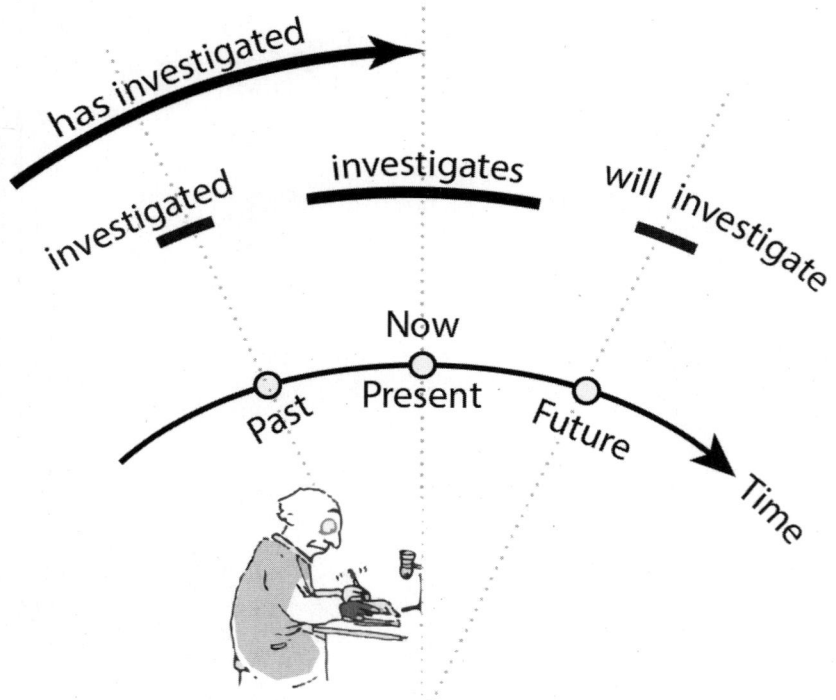

Figure 10.2. Schematic diagram of the time of action of the verb *investigate* in the most common English tenses used in scientific writing. The modal auxiliary is not shown, as it may appear in any time frame.

There are many verb tenses in English to show fine gradations about when something happens. Most scientific writers use few verb tenses and they probably should use even fewer. It is rarely preferable to use a more complex construction when a simpler one would do. The most useful tenses in scientific writing are schematically presented in Figure 10.2 and described below:

- **Past tense** shows completed actions or results that may be specific to the current investigation. Use it to describe apparatus, procedures, and results, and to cite specific works in the literature. Examples are given in Table 10.14.
- **Present perfect tense** (*have or has* + past participle) shows events over an extended period in the past and possibly continuing presently. Use it to cite extensive past work that extended over a prolonged period. Examples are presented in Table 10.15.
- **Present tense** shows action occurring currently or a general truth. Use it to present scientific laws and results that are generally applicable. Examples are presented in Table 10.16.
- **Future tense** shows events that will occur in the future. Use it in research proposals to present proposed methods and the work plan. Examples are given in Table 10.17.
- **Modal auxiliaries** show potential, possibility or doubt. Typical modal auxiliaries add *can, could, may, might, must, ought to, shall, should, will,* and *would* to the present or past participle of the verb. *Might* and *may* are used interchangeably for expressing possibility. However, *might* can be used to express a hypothetical situation. For example, *Nuclear power might not be feasible, without Einstein's groundbreaking work.* Examples are presented in Table 10.18.

Table 10.14. Simple past tense — examples in active voice.

Where used	Example
Statement of purpose — research centered	The objective of the present study <u>was</u> to correlate systolic anomalies with genetic factors.
Restatement of a hypothesis in the Discussion	State-theory <u>predicted</u> that instabilities would be observed above a threshold voltage.
Apparatus — what you used	The radar receiver <u>consisted</u> of four principal parts: an RF amplifier, a mixer, a local oscillator, and a detector.

Where used	Example
Procedure — what you did	The ultrasonic cleaner <u>washed</u> the samples with alcohol prior to mounting.
Results — Presentation (preferable)	However, the entire range of parameters investigated <u>produced</u> the same output voltage.
Discussion — explanation restricted to specific study	The increased coating hardness <u>was</u> the result of increased nitrogen incorporation.
Accomplishments summarized in the discussion/conclusions	This study <u>determined</u> that substrate biasing increased coating adhesion.

Table 10.15. Present perfect tense — examples in active voice.

Where used	Example
Introduction — literature review	Many surgeons <u>have used</u> amputation to treat gangrene in the extremities of diabetes patients.
Introduction — literature review; Results, Discussion	The Nile river <u>has discharged</u> 84×10^9 m^3 of water annually in recent years.
Business Plan — market survey	Glass manufacturers <u>have generated</u> revenues of $75 billion annually.

Table 10.16. Present tense — examples in active voice.

Where used	Example
Research question	How <u>do</u> the dielectric properties of pure epoxy and epoxy/crepe composites <u>differ</u>?
Paper-centered Statement of Purpose — what is done in the paper itself	The objective of this paper <u>is</u> to prove the proposition that EFG filtering <u>optimizes</u> the signal to noise ratio in multi-input receivers with cross-coupled noise.
Sometimes, standard apparatus or what is generally done	Radar receivers <u>process</u> the waves reflected from approaching aircraft.
References to features (e.g. equations, figures, tables) found in the present paper.	Table 3 <u>shows</u> that the acceptance criteria were met only by the laser stabilized feedback loop. Equation 24 <u>suggests</u> that the process efficiency can be improved by operating at higher temperatures.
Results — Location sentence	Table 5 <u>summarizes</u> the composition and wear properties of coatings deposited under various conditions. Fig. 1 <u>shows</u> the dielectric loss factors of pure epoxy resin and epoxy /crepe paper composites at 160°C.

Where used	Example
Results/comment — comparisons	The bit error rate in this case <u>is</u> 5 times greater than that obtained with the Bernstein code. The present results <u>disagree</u> with Smith's earlier findings [23].
Discussion — generally applicable, explanations, implications	The present results <u>demonstrate</u> the necessity of controlling the axial magnetic field. Increased nitrogen incorporation <u>produces</u> harder coatings.

Table 10.17. Future tense — examples in active voice.

Where used	Example
Research report Discussion — future work	The UMC Space Propulsion Laboratory <u>will apply</u> the design principles of the cathode spot thruster developed in this work to the development of a LEOS orbital stabilizer.
Research proposal methodology	The sample holder <u>will bias and heat</u> the sample.
Research proposal work plan	Stage 5 of the proposed program <u>will use</u> the data developed during Stage 4 to optimize the accelerator current.

Table 10.18. Modal auxiliaries — examples in active voice.

Where used	Example
Value statements	The results <u>could facilitate</u> improved long-term monitoring of high voltage bushings.
Results — comment/possible explanations	The less effective error identification routine <u>might explain</u> the increased overhead.
Discussion of results — speculative	The more efficient packing and unpacking operations <u>might be</u> the cause of the increased throughput.
Generalizations	It is likely that placing acoustic absorbers on the partitions between firing positions, as well as on the ceiling and floors, <u>can</u> further reduce the peak pressures.

10.3.1.2 *Verb voice*

The voice of a verb describes the relationship between the action or state that the verb expresses and the participants identified by its arguments (subject, object, etc.). When the subject is the agent or doer of the action, the verb is in the active voice. When the subject is the receiver of the action, the verb is in the passive voice. The active voice is considered more forceful and the usual recommendation in general writing is to avoid the passive voice unless it is really needed. However, in scientific writing, the use of the first person, *I* or *we*, is generally avoided and the passive voice is often preferred.

In Defense of the Passive Voice

Some recent textbooks and some journals favor using a first person subject and the active voice (e.g. <u>*I* maintained</u> *a constant sample temperature by immersing it in a water bath*.). This usage has advantages – it is simple and forceful.

However, most researchers avoid the first person (*we*, and especially *I*) and favor the passive voice (e.g. *A constant sample temperature <u>was maintained</u> by immersing the sample in a water bath*). We strongly advocate the latter because:
- Using *I*, or even *we*, sounds egotistical. Avoiding them helps preserve the researcher's modesty.
- When used repetitively, *I* or *we* is boring.
- Most important, making *I* or *we* the subject of the sentence emphasizes the role of the researcher. In contrast, using the passive voice correctly emphasizes the research.

Examples of the use of active and passive voice in different sections of the research paper are shown in Table 10.19 below.

Table 10.19. Active and passive voice — examples.

Section	Example	Explanation
Apparatus description — human agent: usually passive	*The arc current <u>was controlled</u> by varying the phase angle of the SCR ignition.* [passive]	Emphasizes the receiver of the action rather than the researcher who probably varied the phase angle, e.g. by turning some control knob.

Section	Example	Explanation
Apparatus description — no human agent: either passive or active voice depending on the emphasis	*The video signal was sampled by the frame grabber at a rate of 5 MHz.* [passive]	Emphasizes the receiver of the action.
	The frame grabber sampled the video signal at a rate of 5 MHz. [active]	Emphasizes the role of the equipment or materials.
Procedure — usually passive	*A sequence of pulses with the peak voltages increased in steps of 1 kV was applied to the test gap.* [passive]	Emphasizes the procedure, not who did it.
Results — Location sentences: either passive or active voice may be used	*The dependence of the arc voltage on the gap distance is shown in Figure 1* [passive]	Emphasizes the relationship.
	Figure 1 shows the dependence of the arc voltage on the gap distance. [active]	Emphasizes the figure.
Results — Presentation sentences: either passive or active voice may be used	*Increasing the gap distance linearly increased the arc voltage.* [active]	Emphasizes the gap distance.
	The arc voltage was increased linearly with the gap distance. [passive]	Emphasizes the arc voltage.

The passive voice may be chosen to place old information first in a sentence. Examples are given in Table 10.20:

Table 10.20. Use of passive voice to put old information first in the sentence.

Example	Explanation
The arcs were generated with a welding power supply. The welder produced d.c. current of 0–400 A, with an open circuit voltage of 70 V. The current was terminated when the computer generated enable signal was removed from the supply. The enable signal was terminated with a 50 Ω resistor to prevent reflection.	←passive voice puts arcs, which were presumably already defined, first ← active voice puts known welder first ←passive puts current first (current is assumed to be known) ← enable signal is old information — using passive voice places it first in the sentence

10.3.1.3 *Number*

Verbs are either singular or plural. The verb number must agree with the subject number. "Regular" verbs have standard forms in the various tenses and numbers, and their singular and plural forms differ only in the present tense and third person, which is commonly used in scientific writing. Oddly, although regular plural nouns end in an *s*, regular singular verbs end in an *s*. For example:
- *The <u>oscillator generates</u> a 5 V signal.* [singular]
  The <u>oscillators generate</u> 5 V signals. [plural]
- *<u>Electron velocity increases</u> with temperature.* [singular]
  <u>Electron velocity and electron pressure increase</u> with temperature. [plural]

10.3.1.4 *Commonly used verbs in the physical sciences*

Table 10.21–10.25 list commonly-used verbs together with explanations and examples. The verbs are classified by the sections of the research report in which they typically first appear: Introduction, Methodology, Results, Discussion and Conclusions and the specific part of Conclusions relating to suggested future work. These verbs, however, may appear in other sections and in other genres.

Table 10.21. Verbs commonly used in the Introduction, usually in the past tense. Either the active or passive voice may be appropriate.

Verb	Meaning	Examples
Description of research		
Address	Deal with or discuss, usually with a problem or a matter of concern or controversy.	*Donaldson <u>addressed</u> the issue of double bonds in alkenes.* [active]
Attempt	Try, undertake — usually with the implication that success was not achieved.	*Albert <u>attempted</u> the synthesis of 3-ethyl-2-methylhexane, but he did not obtain sufficiently pure material.* [active]
Consider	Look at or examine — usually a specific factor.	*Smith <u>considered</u> the effect of secondary electron emission on the heat balance of the cathode.* [active]

Verb	Meaning	Examples
Examine	Tends to imply a narrower or possibly more detailed scrutiny of a particular aspect or method than "investigate". May also indicate a cursory review. Often used with some sort of visual examination.	The samples _were examined_ with a scanning electron microscope. [passive] The samples _were examined_ for obvious faults when taken out of the oven. [passive]
Investigate	Conduct research. Similar in meaning but more commonly used than "research".	The influence of the oxygen partial pressure on the deposition composition _was investigated_. [passive] or Smith _investigated_ the influence of the oxygen partial pressure on the deposition composition. [active]
Study	May be used as a synonym for investigate or in a broader sense — to learn a field of knowledge.	Smith _studied_ the impact of the oxygen partial pressure on the deposition composition. [active] The student _studied_ chemistry. [active]
\multicolumn{3}{c}{Description of quantitative aspects of research}		
Analyze	Separate a material into constituent parts or examine carefully to determine key elements (may apply to chemical or mathematical analysis — or even grammatical analysis).	The composition of the compound _was analyzed_ using flame spectroscopy. (passive) or Smith _analyzed_ the composition of the compound using flame spectroscopy. [active]
Evaluate	Determine the significance or value of something. In mathematics — to calculate the numerical value of a formula, function or relation.	The statistical significance of the data _was evaluated_ using XPVC software. [passive]
Measure	Ascertain the extent, quantity or dimensions of something.	The distance to the target _was measured_ interferometrically. [passive]
Quantify	Determine or indicate a measurable quantity. Often used in theoretical work.	Finkelstein _quantified_ the Mg content in slags from 15 ancient sites in the Mediterranean basin. [active]

Verb	Meaning	Examples
Description of research outcome		
Demonstrate	Make evident, show, prove.	Smith _demonstrated_ that the treatment was effective against candida infections. [active]
Determine	Indicate the act or the intent of settling the facts of the matter — reaching a clear conclusion.	Smith _determined_ that oxygen partial pressure caused the deposition composition to.....[active]
Verify	Ascertain the correctness of.	Smith _verified_ Jones's theory. [active]
Referring to reports and publication		
Discuss	Write about, examine the implications.	Smith _discussed_ the economic consequences of introducing the new technology. [active]
Publish	Issue research results in a journal paper or book.	Wright _published_ the first description of vacuum discharge deposition in the 1870's. [active]
Report	Convey research results (in writing or orally).	Edwards _reported_ construction of a Colpitts oscillator operating in the 100 GHz region. [active]
Summarize	Provide a concise overview of a topic. Typically refers to review articles.	Hutchinson _summarized_ progress to date in developing an x-ray laser. [active]

Table 10.22. Verbs commonly used in the methodology section. Usually the past tense and the passive voice are used.

Verb	Meaning	Example
Verbs describing specific action related to process or procedure		
Adapt	Modify, adjust.	Issacson's algorithm _was adapted_ to the present case by including a high-pass filter. [passive]
Adopt	Act in accordance with (e.g. a plan), select, take as one's own.	Industry _adopted_ the 801.11i standard for future units. [active]
Apply	Make use of, put to use for a specific purpose.	Chebyshev filtration _was applied_ to the output signal. [passive]

Verb	Meaning	Example
Assemble	Put or fit together (the parts of).	The test chamber *was assembled*, evacuated, and baked-out. [passive]
Assume	Take for granted, presuppose, postulate.	Local thermodynamic equilibrium *was assumed*. [passive]
Calibrate	Accurately adjust to give correct read-outs (for measurement instruments).	The instrument *was calibrated* using a source traceable to an NIST primary standard. [passive]
Calculate	Determine by mathematical or numerical manipulation.	The required dosage *was calculated* based on NIH recommendations. [passive]
Collect	Gather, make a collection of.	Samples *were collected* of surface water at 8 locations in the Sharon region. [passive]
Construct	Build, erect, set up, assemble.	The hot wire velocimeter *was constructed* from 24 µm diameter nichrome wires. [passive]
Dissect	Cut apart (an animal body, plant, etc.) to examine the structure, relation of parts, or the like; examine minutely part by part; analyze: dissect an idea.	The heart *was dissected* post-mortem, and examined for occurrences of scar tissue. [passive]
Embed	Fix into a surrounding mass, incorporate as an integral part, insert, a set mapped into another set (mathematics).	Iron particles *were embedded* in the silicate matrix. [passive]
Formulate	Reduce to a formula, express in precise form.	The relationship between carbon content and hardness *was formulated* from empirical data. [passive]
Generate	Cause to be, produce.	The 850 GHz signal *was generated* with a travelling wave tube. [passive]
Modify	Make changes, usually minor, typically to improve something.	The standard test protocol *was modified* to accommodate the over-sized samples. [passive]

Verb	Meaning	Example
Mount	Set, fix or place (in apparatus).	The specimen _was mounted_ on the specimen holder. [passive]
Plot	Represent by means of a curve drawn from point to point.	The relationship between hardness and carbon content _is plotted_ in Fig. 2. [passive]
Position	Put in place.	The emitter _was positioned_ using a screw mechanism. [passive]
Simulate	Create a model or likeness of, calculate a result using a model.	The model _simulated_ the effect of carbon content on hardness. [active]
Validated	Substantiate, confirm.	The model _was validated_ by measuring the protuberance-vorticity relationship in a wind tunnel. [passive]
\multicolumn{3}{Verbs relating in a more general sense to process or procedure}		
Carry out	Execute, accomplish, complete.	The measurements of the conductivity _were carried out_ on three samples. [passive]
Conduct	Do, carry out, accomplish.	Heavyside _conducted_ investigations of the electrical properties of the upper atmosphere. [active]
Perform	Do, carry out, accomplish.	Goodyear _performed_ experiments in cross-linking rubber compounds. [active]
Use/Utilize (Use preferred for reasons of simplicity)	Employ for a purpose, put into service, make use of.	A 4-point probe _was used_ to measure surface conductivity. [passive]

n.b. This group of verbs, used in conjunction with noun forms of verbs (e.g. _measurements, investigations, experiments, to measure_) is overused. Much stronger, terser sentences can be written using the original verb forms:
The conductivity of three samples was measured.
Heavyside investigated the electrical properties of the upper atmosphere.
Goodyear experimented with cross-linking rubber compounds.
The surface conductivity was measured with a 4-point probe.

Table 10.23. Verbs commonly used in reporting results, usually in the past tense. Either the active or passive voice may be appropriate.

Verb	Meaning	Examples
Verbs used to present results, usually in the general sense		
Achieve	Obtain a desired result or objective.	The performance specifications _were achieved_ using the CF32 processor. [passive]
Observe	Notice, usually something significant.	Blue light emission _was observed_ from the sample during electron bombardment. [active]
Obtain	Get, acquire.	The clearest images _were obtained_ with 25 μs exposure. [passive]
Verbs expressing variation		
Decrease	Become or make less in size, amount, intensity or degree.	The thermal radiation to the sample _was decreased_ by insertion of a heat shield [passive]
Decline	Become lower in amount or less in number, worsen in condition or quality.	Toxicity _declined_ with the addition of $MgCl_2$ to the solution. [active]
Grow	Increase in size — usually relates to a living thing.	The coating thickness _grew_ slower when the surface was bombarded with Ar ions. [active]
Increase	Become or make greater in size, amount, intensity or degree.	The voltage _increased_ exponentially as a function of the sample temperature. [active]
Maximize	Make as large possible, make the best use of.	The hardness _was maximized_ when the samples were heat treated at 900 °C for 10 min, and then forced-air cooled. [passive]
Minimize	Reduce to the smallest amount possible.	Energy consumption _was minimized_ by disabling the CPU and disk drive during stand-by. [passive]
Oscillate	Vary periodically; move or swing back and forth at a regular speed.	The output voltage _oscillated_ at a frequency that increased with the bias current. [active]
Reach	Increase up to a certain level.	The current _reached_ its maximum value of 23.4 kA at 32 μs after pulse initiation. [active]

Verb	Meaning	Examples
	Verbs used to present specific results	
Derive	Reach by mathematical manipulation, deduction, or reasoning; obtain something from a specific source.	An expression <u>was derived</u> for the particle velocity based on conservation of momentum, conservation of energy, and Maxwell's equations. [passive] Digitalis, which is <u>derived</u> from the foxglove plant, is used to treat congestive heart failure. [passive]
Identify	Establish what is.	The presence of hydroxyl free radicals <u>was identified</u> in the solution after arc treatment. [passive]
Recognize	Identify.	After edge enhancement, facial expressions could be easily <u>recognized</u> in the tested photographs. [passive]
Solve	Find an answer or explanation for a problem or equation.	The set of equations <u>was solved</u> numerically using Runga-Kutta integration. [passive]
	Verbs used to express causality	
Cause	Give rise to an action or condition, make something happen	Ion bombardment <u>caused</u> internal collision cascades. [active] Internal collision cascades <u>were caused</u> by ion bombardment. [passive]
Influence	Have an impact on something	The background temperature strongly <u>influenced</u> the results. [active]
Produce	Make or manufacture from components; cause a particular result.	Fenton's reaction <u>produced</u> hydroxyl radicals in the solution. [active]

Table 10.24. Verbs commonly used in the Discussion and Conclusions — usually in past tense and active voice but depending on degree of certainty may be in present tense.

Verb	Meaning	Examples
	Verbs used in presenting general conclusions	
Indicate	Point out or show, suggest as a an explanation or a desirable or feasible course of action.	The observation of ruptured membranes as well as the decreased number of colony forming units <u>indicated</u> that the plasma treatment effectively killed the bacteria. [active]

Verb	Meaning	Examples
Provide (insight into)	Help to understand — *provide insight into* is often overused	*The combination of video observation and emission spectroscopy <u>provided insight into</u> the plasma dynamics during high current arcing.* [active]
Suggest	Imply, hint, put forward a result or explanation that is well based, but not proven	*Ellipsometric measurements of the monolayers on the Ge surface <u>suggest</u> that the alkane chains were dense and well formed, and tilted less than 10° with respect to the surface.* [active]
Verbs used in presenting comparisons		
Compare	Note the similarity or dissimilarity between	*The results <u>were compared</u> and found to be identical.* [passive]
Contrast	Note the dissimilarity.	*The present results using low-voltage pulses strongly <u>contrasted</u> with Smith's earlier results that used high-voltage.* [active]
Verbs used in presenting value of the research (may be in future tense)		
Enable	Provide the means to do something. make possible.	*The image processing algorithm developed in this work <u>will enable</u> the incorporation of focus correction into miniature cameras, e.g. in smartphones.* [active]
Facilitate	Made more feasible or easier.	*This research <u>will facilitate</u> development of a grey-water processor for domestic washing machines.* [active]
Help	Aid, assist. make easier.	*Widespread improvement in sanitary conditions <u>will help</u> eradicate malaria in tropical regions.* [active]
Improve	Make or become better	*Edge enhancement <u>improved</u> the feature recognition in the images.* [active]
Verbs used to indicate support/confirmation of a claim, previous work or theory		
Confirm	Establish the truth or correctness of something, usually previously believed to be the case.	*The high resolution images <u>confirmed</u> Harrison's supposition that the impurities were concentrated at the grain boundaries.* [active]

Verb	Meaning	Examples
Establish	Set up, institute, bring about but also to fix or to confirm.	Uganda *established* a science and technology coordination council for focusing funding to projects critically needed for its economic development. [active]
Prove	Demonstrate the truth of. [*Prove* is a very strong term and should be used cautiously.]	*The results reported here prove that Copaxone is effective in treating Secondary Progressive Multiple Sclerosis.* [active]
Support	Suggest the truth of, corroborate.	*The results obtained in this study support Richardson's theory.* [active]
Verbs used to indicate rejection/disagreement with a claim, previous work or theory		
Contradict	Disagree, deny the truth of [usually of something previously believed to be correct]; assert the opposite of.	*The results obtained in this study contradict Smith's earlier findings.* [active]
Differ (from)	Be unlike or dissimilar, disagree.	*The response of the coated and implanted samples to a humid environment differed considerably.* [active]
Eliminate	Exclude from consideration.	*Insulating the test specimen eliminated conductive heat transfer to it.* [active]
Refute	Disprove, prove to be wrong or false, deny or contradict. [*Refute* is a very strong term and should be used cautiously.]	*Thus the present results refute Obsolem's model.* [active]
Rule out	Exclude, eliminate.	*Because the same results were obtained both with and without insulating the test specimen, conductive heat transfer may be ruled out as the cause for the observed phase transformation.* [active]

Table 10.25. Verbs referring to future work, usually in future tense or using a modal auxiliary. Either passive or active voice may be appropriate.

Verb	Meaning	*Examples*
Remain (to)	Is left (to do).	*Although XFS was applied successfully to phyllosilicates, tectosilicates <u>remain to be examined</u>.* [active]
Extended	Carry further, make larger.	*The research reported here <u>should be extended</u> to include oxides and sulfides.* [passive]
Should be explored	Call for further investigation of the subject by the scientific community.	*The application of Penicillin to gram-negative bacteria <u>should be explored</u>.* [passive]
Will be explored	Indicate that the researcher intends to further investigate the subject. This may be intended as a hands off signal to other researchers but it would be inadvisable to count on it.	*The application of Penicillin to gram-negative bacteria <u>will be explored</u> in future work.* [passive]

10.3.2 *Nouns and noun phrases*

A noun is a word for a person, place, thing, activity, quality or idea. A noun or noun phrase (or pronoun — see section 10.3.3.) is the subject (explicit or implicit) of the sentence. Noun and noun phrases may also be direct or indirect objects. In English, nouns have the same form, whatever their use in the sentence.

A noun phrase is made up of a central (head) noun and all its modifiers. The modifiers define the noun in greater detail, sometimes in very specific detail. Pre-modifiers (appearing before the noun) may be articles, adjectives or possessive pronouns. Post-modifiers (appearing after the noun) are typically prepositional phrases. Noun phrases can become very complicated. Often, they are necessary to describe the noun specifically. They may greatly enrich language. However, sometimes, they are used because of lack of knowledge of the appropriate term. This is a pitfall especially for non-native speakers.

A compound noun can be composed of several nouns (e.g. electron velocity, ball-point pen, fuel plant). A compound noun expresses a single concept. Functionally, the first nouns in some ways act as adjectives because they describe, limit, or modify the final noun in the compound noun. The compound noun may either be hyphenated, or not. A compound noun (e.g. *cathode surface*) is preferable to using the possessive in relation to an inanimate object (e.g. do **not** use *cathode's surface*). All of the elements of the compound noun, except for the last element, i.e. the noun that is modified, are always used in the singular form, e.g. *electron velocity* and **not** *electrons velocity* (no matter how many electrons are involved).

Some nouns are derived from verbs:
- The gerund (present participle e.g. *exploring, annealing, coating*) functions as a verb when it is used as part of the various progressive tenses. Standing alone, it may function as a noun or as an adjective.
- The infinitive comprises "to" plus a verb (e.g. *to measure, to analyze*), and it may function as a noun.

Some examples of sentences using gerunds and infinitives follow in Table 10.26.

Table 10.26. Gerund and infinitive examples.

Gerund	Infinitive	Explanation
<u>Desiccating</u> the sample allowed examination in an electron microscope	<u>To maximize</u> power is necessary for meeting the specifications.	Subject [gerund more common, bare infinitive form rarely used in scientific writing]
A specific combination of gas direction and speed produced the <u>whistling</u> heard in the cavity.	The purpose of this procedure was <u>to</u> <u>functionalize</u> the nanoparticles.	Direct object [bare infinitive rarely used in scientific writing]
The function of the circuit breaker is <u>protecting</u> the circuit.	The function of the circuit breaker is <u>to</u> <u>protect</u> the circuit.	Subject complement [in scientific writing, the gerund or infinitive would most likely have an object, e.g. ...*the circuit*]

Gerund	Infinitive	Explanation
A digital multi-meter was used for *measuring* the bias potential. A traceable standard is required for *calibrating* the oscilloscope. The concentration was controlled by *diluting* the original solution as needed.	Rare — only after *except* and *but*: The scientist had no choice but *to use* the digital multi-meter for measuring the bias potential	Object of Preposition Caution: do not confuse the infinitive with prepositional phrases beginning with "to"

The gerund and the infinitive forms are not interchangeable when they are used as direct objects. Some verbs may be followed by either, but many may only be followed by one type. A gerund often describes an event in progress, an infinitive often suggests an event in the future. See Table 10.27 below for examples using verbs common in scientific writing.

Table 10.27. Examples of gerunds and infinitives used as direct objects.

Verbs followed by direct object gerunds	Example
Avoid	*Avoid applying* excess voltage.
Consider	Theoreticians should *consider preparing* lecture slides without any equations.
Delay	The buffer *delayed activating* the comparator until the reference signal was received.
Recommend	The international committee *recommended adopting* the new standard.
Suggest	Harrison *suggested implementing* the new process under rigid feedback control.

Verbs followed by direct object infinitives	Example
Agree	The appeals committee agreed *to reconsider* its decision.
Attempt	Harris and Jenkins attempted *to distill* the mixture at medium temperature.
Decide	The admissions committee decided *to reject* the application.
Expect	The subjects were expected *to achieve* their goal in under 5 minutes.
Intend	Each group intends *to finish* their assigned task by Month 9 of the project.
Plan	The committee plans *to find* alternative stimulants to the economy.

Note that insertion of a word between *to* and the verb (*e.g.* to *quickly* ionize, to *slowly* mix) forms a split infinitive. Split infinitives tend to be clumsy and unnecessary. Avoid them.

Typically, plural nouns will end in *s* or *es*, although this is not always the case. Some common exceptions include words that follow Latin forms (e.g. datum/data, ovum/ovae) and some words in which internal vowels change to form the plural (*mouse/mice, goose/geese*).

If the subject noun or noun phrase is plural, the verb also must be in the plural form. When the subject is a noun phrase, subject/verb agreement may be tricky. The verb must agree with the head noun. For example: *The influence* [subject, singular] *of voltage and temperature was examined.* [verb, singular]. Uncountable nouns such as *research, literature, equipment* and *evidence* do not have a plural and take only the singular form of the verb.

10.3.3 *Pronouns and demonstratives*

Pronouns (e.g. *he, she, it, they*) and demonstratives (also called pointing words e.g. *this, that, these, those*) are used to refer to previously mentioned or old information. They can concisely refer back to complex information, thus avoiding repetition. They orient the reader and provide continuity from one sentence or paragraph to another. *It* or *its* (singular) and *they* or *their* (plural) are the most common pronouns used in scientific writing. Note that the neuter singular possessive pronoun (*its*) is written without an apostrophe (e.g. *its color*). *It's* is a contraction of *it is* and is best spelled out in scientific writing.

Pronouns and demonstratives must be used with caution — it must be perfectly clear to the reader to which nouns or old information they refer. Usually the old information should not be more distant than one (or maybe two) sentences.

- Good example: *Vacuum arc coatings are increasingly used in the tool industry. Their deposition rate is much higher than that of most other technologies, thus the economics are favorable. Their* clearly refers to *coatings* in the previous sentence.
- Bad example: *Falcons include kestrels, merlins, hobbies, and especially peregrines. They kill their prey with their beaks, using a tooth on its side. They, their and its* are ambiguous. Do the pronouns refer to Peregrines, to the entire list, or to Falcons in general?

While English nouns are not declined (i.e. change form depending on their case or use in a sentence), pronouns do have different forms for different functions. (Examples: *The samples were washed and dried. Then they* [subject] *were mounted on a test fixture. Their* [possessive] *thermal conductivity was determined first by irradiating them* [direct object] *with a uniform light beam of known fluence, and then measuring the temperature difference between their* [possessive] *front and back surfaces.*

Pronouns are the only part of speech that is commonly differentiated by gender in English. Traditionally, the masculine form was used to express both genders collectively. It is grammatically correct to say *Each student must submit his exercises by Monday* when the students may be male or female, but it is no longer politically correct. Moreover, it seems ever more inappropriate as the numbers of female scientists and engineers grow. The alternatives are cumbersome. *Each student must submit his or her exercises by Monday* or *All students must submit their exercises by Monday* or *You must submit your exercises by Monday* or *Exercises must be submitted by Monday. His or her* is awkward, particularly in longer texts. The plural pronoun *their* is not always appropriate in context and may be confusing. And *you* may seem too personal and directive. One may alternate the use of *he* and *she* by chapters but the result sounds artificial. Switching to passive voice is increasingly more common. Nonetheless, all of these alternatives are preferable to the automatic use of the universal masculine form.

10.3.4 Articles — use of a, an and the

The use of the definite and indefinite articles (*the, a,* and *an*) is particularly confusing for non-native speakers of English. Many languages such as Russian do not use articles, and many other languages do not make a distinction between them in the same way as in English. Even native speakers may be confused at times because article use depends on whether the noun is countable or not, whether the noun is singular or plural, and what information is known to the reader about the noun.

Articles are in the category of "determiners". A determiner appears before a noun or at the beginning of a noun phrase. It tells us whether the noun phrase is general or specific. General determiners are the indefinite articles *a/an,* as well as *any, another, other,* and *what.* Specific determiners are the definite article *the,* as well as possessives *(my, your, his, her, its, our, their, whose),* demonstratives *(this, that, these, those)* and the interrogative *(which).*

A countable noun is one that has both a singular and plural form. It makes sense to ask the question "*how many*" about a countable noun. For example: *How many experiments did the researcher perform?* or "*How many journals are accessible through open access?* Uncountable nouns such as *research, literature, equipment* and *evidence* do not have a plural. It makes sense to ask the question "how much" about a uncountable noun. For example: *How much of the literature did you read?* or *How much evidence is needed to refute this claim?*

Some nouns that are commonly considered uncountable like economy or industry may be countable nouns when they are used to refer to a specific member of the class. For example, industry "may" refer to a specific countable set of industries or may be a generic term. For example: *Many industries use this application* (countable) as opposed to *Industry commonly uses this application* (uncountable or generic).

The most important rules for using *a/an* or *the* in scientific writing are summarized below:

- If the noun is both singular and countable, either the indefinite article *a* or *an* or the definite article *the* or another determiner *(e.g. your, its, this, that* or *a number)* must modify the noun. The noun cannot stand alone without an article or determiner. For example:
<u>A</u> *particular shape may be desired for <u>a</u> given application.*
<u>An</u> *interpolation code filled in picture details which were lost in transmission.*
<u>This</u> *generator will trigger <u>an</u> explosion.*
<u>The</u> *procedure produced <u>one</u> viable specimen.* <u>Its</u> *tail was 3 cm long.*
- The choice of *a* or *an* depends on the sound of the first letter in the word following. If the sound is a vowel, *an* is used, if a consonant, *a*. Sound, not spelling, is the determining factor, so we write <u>an</u> *MRI scan* because the initial sound is *em*.
- If the noun is plural and countable, the definite article *the* or another specific determiner *(e.g. your, their, these, those)* is used. The indefinite article never modifies plural nouns. For example:
<u>The</u> *popular and universally used coating systems* or <u>Those</u> *popular and universally used coating systems.* [Not *a ...systems*]
<u>The</u> *desired properties* or <u>Their</u> *desired properties... [*Not *a desired properties]*
- If the noun is uncountable or generic, no article is used. For example:
Physical vapor deposition is increasingly used for hard film fabrication to improve wear behavior or corrosion resistance.
- The definite article may be used with singular or plural nouns, whether countable or not, when referring to assumed or old information or when the specific meaning is clarified immediately thereafter in the sentence.
    - In scientific writing, the indefinite article or no article (with a generic term) is used the first time a new term is introduced; thereafter the definite article *the* is used in every subsequent mention of the term. The definite article *the* is used to point to old information and tells the reader that this term is old information. For example: *Ion bombardment usually requires the use of <u>an</u> ion source. <u>The</u> source may be of several designs, including the Kaufmann source, and the dualplasmatron. An was used to introduce the ion source. The*

was used to refer to the source in the second sentence because the reader already knows about the ion source from the previous sentence. Note that the noun *ion* was used as part of the compound noun *ion source* in the first sentence, and thus served to modify or describe the word *source*. *Ion* was dropped in the repetition, since the reader already knows what source is being described from the previous sentence. If no other sources are used in the paper, after the first description of the source, the modifier *ion* should not be repeated. This makes the writing more concise and less confusing. If an *ion* source were to appear again later in the paper, the reader would be wondering if there was some other source whose mention he missed.

o The definite article may also point forward to specific information which follows immediately, e.g. *The code generated by Harrison's algorithm has a lower bit error rate than conventional encryption methods*. Here the specification of the *code* (*generated by Harrison's algorithm*) closely follows *The code*.

o The definite article may be used on first mention if the part is obviously known (i.e. it is part of the knowledge that the reader may be assumed to share). For example: *The anode was water cooled and constructed from OFHC copper* would be appropriate if it may be assumed that the reader knows that there is an anode from the scientific context.

10.3.5 *Adjectives*

Adjectives are words that describe or modify a noun. They usually answer the questions: Which one? What kind of? How many? Typically, adjectives are used more sparsely in scientific writing than in general literature. Quantitative measures are more common than qualitative descriptions. The quantitative properties of apparatus and methods are relevant to replicating the results. The material used (e.g. steel, copper, etc.), and quantitative measures such as dimensions, voltage, strength, etc. are likely to be relevant. Information conveyed by qualitative adjectives (e.g. color, big, small, fast, slow) is usually meaningless as it does not allow replication. The extent to which a result is *innovative* or *interesting* should be shown concretely in the presentation of the results and the discussion. It should not be described in general terms. A *significant* result has the specific meaning of being statistically significant.

However, qualitative comparisons are common and may be coupled for emphasis with quantitative comparisons. The comparative (inflected suffixes *-er,* or *-ier* when an adjective ends in y, or *more*) is used for comparing two things and the superlative (*-est* or *-iest* or *-most*) for comparing three or more things. Generally, the inflected suffixes are used for adjectives of one or two syllables. *"-more"* or *"-most"* are always used with adjectives of more than two syllables. The word *than* frequently accompanies the comparative and *the* precedes the superlative, e.g. *Aluminum is a strong material. However, titanium is stronger than aluminum. Among the materials tested, the titanium based super alloy was the strongest.*

When using a comparative adjective, the basis of comparison must be clear to the reader, either from context, or by providing the comparison with a *than* statement. *The anodic reaction was fast. However, the cathodic reaction was faster.* [The basis of comparison for the comparative *faster* is clear in the context of the first sentence.] *The cathodic reaction was faster than the anodic reaction. The Cu-Al alloy was stronger than the Cu-Sn alloy.* [The basis of comparison is specified by the *than* phrase.]

The expression *in comparison to* should not be used together with *than*. It is incorrect to write: *The Cu-Al alloy was stronger than in comparison to the Cu-Sn alloy.* Moreover, it is generally preferable to use *than* in preference to *in comparison to*. Instead of: *However, titanium is stronger in comparison to aluminum,* it is preferable to write *However, titanium is stronger than aluminum.*

Be careful not to use *more* with a comparative adjective formed with *-er* nor to use *most* with a superlative adjective formed with *-est* (e.g., do not write that something is *more heavier* or *most heaviest*). Avoid forming comparatives or superlatives of adjectives that already express an extreme of comparison, e.g. *absolute, sufficient, unavoidable, unique,* and *universal*. *"More absolute"* is not possible.

Some commonly used adjectives have irregular forms in the comparative and superlative degrees, as shown in Table 10.28:

Table 10.28. Irregular comparative and superlative adjectives.

Adjective	Comparative	Superlative
Good	Better	Best
Bad	Worse	Worst
Little	Less	Least
Much/many/some	More	Most

When comparing quantities, *fewer* is generally used for things that are countable while *less* is used for measurable, uncountable quantities. *In the second case, there were fewer excited atoms* [countable] *and their energy* [measurable] *was less. Less* is also used for mathematical expressions in the sense of *minus*, e.g. *The cooling system must remove all of the combustion power, less the power delivered to the drive train.*

Adjectives nearly always appear immediately before the noun that they modify. There are a few exceptions. When indefinite pronouns such as *something, someone* or *anybody* are modified by an adjective, the adjective comes after the pronoun, e.g. *something strange happened.*

Sometimes, a string of adjectives will be used. Order is usually intuitive for the native speaker, but the rules tend to be complicated. Speakers of English as a second language often have difficulty with adjective order. Michael Swan (Practical English Usage, Oxford University Press, 1997; referenced in BritishCouncil.org) lists some of the most important rules which may be summarized as follows; **(article) + number + judgement/attitude + size, length, height + age + color + origin + material + purpose + noun.** For example: *The fourteen successful 24 cm Cu samples were further tested for durability* and *The eight undamaged 12 cm diameter prehistoric blue Sicilian bronze drinking cups were displayed in the museum.*

Note that modifier nouns—nouns used as adjectives, e.g. *electron velocity* — together with the noun that is modified form a compound noun that must stay together. All of the real adjectives must precede all elements of the compound noun.

It is not advisable to use more than two or three adjectives together. When adjectives belong to the same category, a comma is placed between them but not between the last adjective and the noun: *the polished, extruded, aluminum nozzle.*

Adjectives that are really participles, i.e. verb forms with *-ing* and *-ed* endings, are easily confused. Generally, the *-ed* ending means that the noun so described has a passive relationship with the noun it modifies. For example: *Interested parties include government, industry and university nano-fabrication facilities.* or *The ionized atoms were detected with a negatively biased probe.* The *-ing* ending often means that the noun described has a more active role. For example: *The experiment produced interesting results.* or *The specimens were protected from ionizing radiation by a 3 cm thick lead shield.*

10.3.6 *Adverbs*

Adverbs modify verbs, adjectives or other adverbs. They often indicate when, where, why or under what conditions something happens. They also tell us to what extent or how much. Adverbs frequently end in *-ly (e.g. deliberately, dramatically, exactly, hourly, occasionally, quickly, perfectly, powerfully, rapidly, rarely, regularly, safely, sharply, slowly, technically)*; however not always *(e.g. always, often, seldom, sometimes, very, well)*. An *-ly* ending is not a guarantee that a word is an adverb (e.g. *costly, early, only, orderly, smelly,* and *unlikely* are adjectives). Like adjectives, adverbs can have comparative and superlative forms to show degree. *More* and *most* or *less* and *least* are used to show degree with adverbs.

Adverbs do not have a fixed place with respect to the words that they modify. For example, both *"Oxygen strongly bonded with the carbon atoms"* and *"Oxygen bonded strongly with the carbon atoms"* are correct.

If there are more than two or three adverb modifiers, particularly when they are prepositional phrases, it is typical to move one or more to the beginning of the sentence e.g. [*Before being weighed, the samples were thoroughly dried in a tube furnace.*] rather than [*The samples were thoroughly dried in a tube furnace before being weighed.*] Putting the adverbial modifier at the beginning of the sentence emphasizes that modifier.

Inappropriate adverb order can lead to some strange sentences. For example: *The professor planned the organization of the lab in the shower.* Is the lab really going to be located in the shower? It would be clearer to say: *While the professor showered, he planned the organization of the lab.* Word processing grammar checkers rarely catch this type of error.

Avoid using adverbs such as *really, very, quite, extremely* and *severely* in scientific writing. Instead, try to be quantitative, and allow the reader to conclude based on the data whether an action was extreme or not.

10.3.7 *Conjunctions*

A **coordinating conjunction** connects serial lists or independent clauses in compound sentences. In scientific writing, the most commonly used coordinating conjunctions are *and*, *but*, and *or*. Each has a different meaning as summarized in Table 10.29.

Table 10.29. Coordinating Conjunctions.

Conjunction	Use to:	Example
And	Express chronological sequence	*The samples were weighed, sorted and mounted.*
	Indicate that an action is the result of another action	*The trigger signal generated a high voltage pulse and subsequently an explosion.*
	Connect a series of related items	*Bacteria, fungi and viruses were found in the contaminated water sample.*

Conjunction	Use to:	Example
But	Connect two ideas with the meaning of the exception of	The voltage increased with time in most of the samples, <u>but</u> it decreased in the alumina sample.
	Connect two ideas that are not in agreement	All of the transistors passed initial inspection, <u>but</u> nonetheless 18% of the transistors failed during the first hour of burn-in testing.
Or	Define alternatives	Each ovum was categorized as viable <u>or</u> non-viable.
	To suggest that one possibility can be realized but not the other	The components must be tempered, <u>or</u> the assembly will fail quickly during service.

Despite what many of us learned in English classes, coordinating conjunctions may be used to begin sentences when the writer wishes to emphasis a transition from the previous sentence. If the two sentences are not overly long or complex, it may be preferable to combine them into one compound sentence or use a semicolon to separate them. Sometimes the transition is clear without the use of the initial conjunction, and then it may be preferable to omit it.

Correlative conjunctions are used in pairs to join sentence elements that are of equal importance. Examples are given in Table 10.30.

Table 10.30. Correlative Conjunctions.

Conjunction	Use to	Example
Both….and	Connect two related items of similar importance.	<u>Both</u> gram-positive <u>and</u> gram-negative bacteria were tested.
Not only….but also	Connect two related items when the first is more important than the second.	<u>Not only</u> did the ClearWave window have better visibility, <u>but</u> it <u>also</u> had lower microwave leakage.

Conjunction	Use to	Example
Either....or	Connect two alternatives, where either may be applicable.	Electron density may be measured *either* by interferometry *or* by using a Langmuir probe.
Neither...nor	Connect two alternatives where neither are applicable.	*Neither* alloying *nor* annealing produced the required ductility.
Whether .....or	Connect two alternatives where the result is similar.	The frictional wear was too large for field deployment *whether* the gear wheel was heat treated *or* coated.

A **subordinating conjunction** introduces a subordinate (or dependent) clause and establishes the relationship between the dependent clause and the rest of the sentence. It turns the clause into something that depends on the rest of the sentence for its meaning. Common subordinating conjunctions are listed in Table 10.31.

Table 10.31. Common Subordinating Conjunctions.

Conjunction	Use to express	Example
As	Comparison. Compare similar items, to introduce an example or to indicate "in the manner of".	The tempered alloy samples were *as* ductile as the annealed samples. Glass microscope slides were used *as* deposition substrates.
As if As though	Compares to an unlikely situation.	The experiments were conducted *as if (as though)* the previous results were not already known.
Than	Introduces the second element in an unequal comparison, an alternative or in expressions of preference, a rejected choice.	The measured tensile strength was 50% greater *than* required.
Rather than	Instead of	*Rather than* ethanol, methanol was used to dilute the reagent.

Conjunction	Use to express	Example
	Condition	
If	On condition that, even though	The voltage exceeded 21 V <u>if</u> the gain was at least 12.
Where	In, at or to what place, position or circumstances	The samples were polished <u>where</u> needed to remove surface oxidation.
Wherever	In, at, or to whatever place, in any condition	Imidacloprid was applied <u>wherever</u> aphids were detected.
	Contrast	
Although	Despite the fact that or even though	<u>Although</u> the Al samples withstood the corrosive atmosphere, all of the Fe samples were corroded.
Unless	Except if	The samples were stored for subsequent analysis, <u>unless</u> they were discolored.
Whereas	In view of the fact that	<u>Whereas</u> a surface hardness of at least 42 GPa was required to protect the surface, a superhard Si-Ti-N coating was deposited.
While	At the same time as. It is often misused instead of *although* or *whereas*.	The bacterial sample for the next run was prepared <u>while</u> the bacterial growth was measured for one run.

Conjunction	Use to express	Example
Because	**Cause, Reason** The cause or reason.	*The theoretical predictions were less than the measured values <u>because</u> secondary emission was neglected.*
In order that, so that	A precondition in order for something to happen.	*Heating to 43°C was required <u>in order that (so that)</u> the material would be fully coagulated.*
Since	May indicate in view of the fact that or because, but more commonly used in a temporal sense in relation to a time in the past.	*<u>Since</u> the dawn of the Industrial Revolution, steel has fulfilled an important role in materials engineering.*
After Before	**Time sequence** x follows y x precedes y	*<u>After</u> drying, the samples were mounted.* *<u>Before</u> mounting, the samples were dried.*
Once	**Time relationship** Following which	*The samples were mounted <u>once</u> they were dried.*
Then	Next in time, space or order	*First the samples were dried, <u>then</u> they were mounted.*
Until	Up to	*The samples were stored in a storage closet <u>until</u> they were completely dry.*
When	At the time	*<u>When</u> the sample moisture reached 23%, the sample was removed from the drying chamber and mounted in the test chamber.*
Whenever	At any time	*<u>Whenever</u> the moisture of the sample was 23%, it was removed from the drying chamber and mounted in the test chamber.*

The relative pronouns *that, which* and *who* also introduce dependent clauses but the pronoun itself acts as the subject of the dependent clause whereas the subordinating conjunction does not. *That* and *which* refer to things, *who* to human beings. *That* or *who* introduce restrictive clauses, i.e. essential phrases needed to complete the meaning of the sentence. For example: *The samples that were contaminated during pre-treatment were removed from the experimental batch. Only subjects who passed the physical examination were subjected to the high stress stimuli.* *Which* or *who* introduce clauses providing additional but nonessential information (i.e. the sentence would make sense without the additional information). These clauses are called non-restrictive clauses (or non-essential clauses) and are bracketed by commas For example: *The magnetometer measurements, which were conducted at the Weizmann Institute, showed that the nanoparticles were super-paramagnetic. The subjects, who were all recruited by advertising on campus, fasted overnight before preliminary blood testing.*

10.3.8 *Sentence connectors*

Sentence connectors are used to express the relationship between ideas. They give paragraphs coherence. Sentence connectors are usually placed at the beginning of a sentence. Table 10.32 describes useful connectors.

Table 10.32. Sentence connectors.

Connector	Use to express	Example
First, second, third, etc. Next, last, finally	Sequential order	*The sample preparation consisted of three stages. First, the samples were dried in hot air. Second, they were annealed at 350°C for 10 min. Third, they were mounted with epoxy in the sample chamber.*
Also / Furthermore / In addition	Logical order	*In addition to the normal procedure, an extra mechanical cleaning stage was added if the samples were severely corroded.*
Most / more importantly Most significantly Above all Primarily	Order of importance	*More importantly, it was found that AJED processed samples performed better than all of the other samples. Above all, they were stiffer and harder. Most significantly, their lifetime was 50% greater than any of the other samples.*

Connector	Use to express	Example
As a result As a consequence Consequently	Result	*As a result* of AJED treatment of the gear wheel, the overall expected lifetime of the vehicle could be increased by 28%.
Therefore Thus	*Therefore* and *thus* indicate that the following sentence or phrase is an inescapable logical conclusion which must follow from the proceeding material. If this is not the case, e.g if additional information is needed to reach the conclusion, then another connector should be used.	The magnetic nanoparticles were produced in an arc discharge between pure graphite electrodes submerged in pure ethanol. SIMMS analysis found only C, and specifically no metallic contaminants, in the nanoparticles. *Therefore*, the observed magnetism came from the carbon.
The reason for	Causality	*The reason for* the superior performance was reduced friction of wear parts.
Similarly Likewise	Comparison of similar items or concepts.	*Similarly*, wear of these parts was decreased.
However On the other hand On the contrary In contrast	Contrast	*However*, the ion implanted samples were too brittle, and failed under impact testing.

10.3.9 *Prepositions*

Prepositions are short words commonly used to show a relationship in space, time or logic between two or more people, places or things. A preposition is usually part of a structure called a prepositional phrase. Prepositional phrases are generally constructed with a preposition followed by a determiner and an adjective or two that are followed by a pronoun or noun (called the object of the preposition). For example:

between [preposition] *the* [determiner] *energetic* [adjective] *ion source* [compound noun] *and the spherical* [adjective] *target* [noun]. This whole phrase takes on a modifying role, by locating something in time and space, describing a noun, or telling when or where or under what conditions something happened.

The translation of prepositions between languages is rarely one to one, and most speakers of English as a second language find choosing the right preposition difficult.

10.3.9.1 *Spatial Prepositions commonly used in scientific writing*

Spatial prepositions express location. A diagram of some spatial prepositions commonly used in scientific writing is shown in Fig. 10.3 below. Please note that some of these prepositions are used in a more general sense, in particular *in, on* and *to* — see Table 10.33.

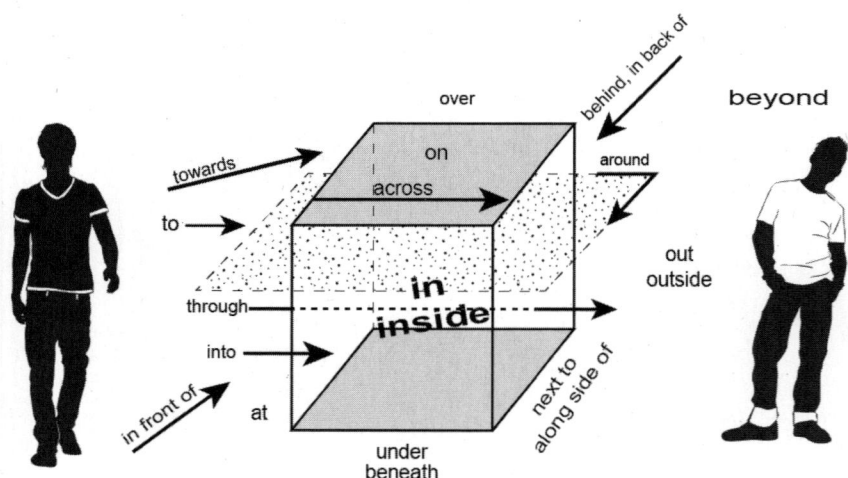

Figure 10.3. Spatial prepositions, indicating location with respect to a box, and seen by an observer.

10.3.9.2 *Temporal Prepositions commonly used in scientific writing*

Temporal prepositions express time. Some temporal prepositions commonly used in scientific writing are illustrated schematically in Fig. 10.4 below.

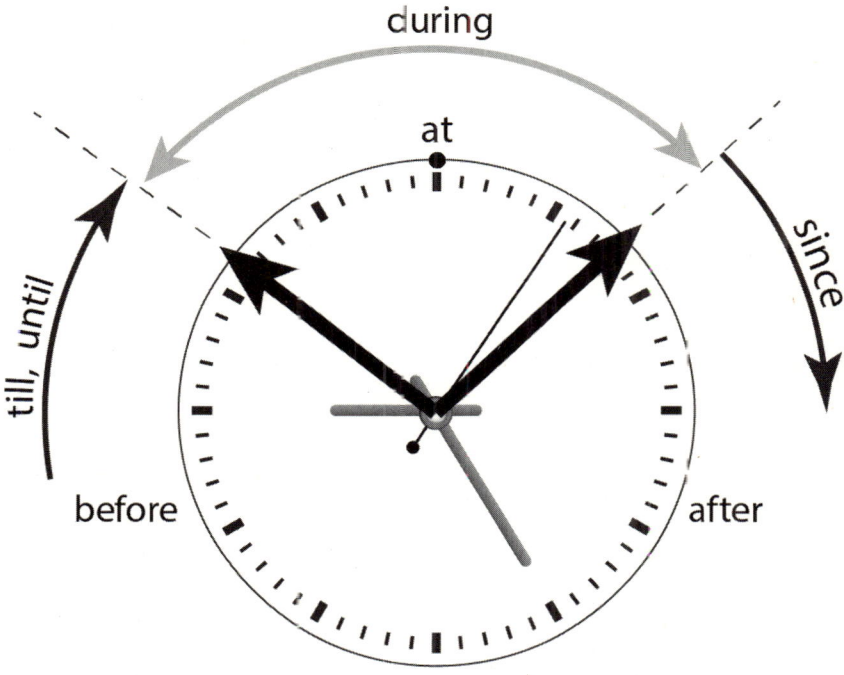

Figure 10.4. Temporal prepositions illustrated with respect to a designated time interval or specified time point.

10.3.9.3 *Other Prepositions*

Prepositions can also express a more abstract, conceptual relationship than time or space. Many of the spatial and temporal prepositions listed in the preceding sections are also used with a conceptual meaning. Table 10.33 gives examples of conceptual prepositions that are commonly used in scientific writing.

Table 10.33. Other prepositions.

Preposition	Expresses	Example
About	On the subject of, concerning	Smith's work was <u>about</u> the effect of magnetic fields on hornets.
Against	Opposition or contrast	The lack of coherence in the output wave argues <u>against</u> lasing in the medium.
By	Means	The surface resistivity was measured <u>by</u> a 4-point probe.
Except	Exclusion	All of the amplifiers exceeded a gain-bandwidth product of 100 GHz <u>except</u> the travelling wave tube circuit.
For	Purpose	A personal computer equipped with an A/D input card was used <u>for</u> recording the data.
From	Indicates starting point of a physical movement or an activity or event; may relate to a source, cause or basis; differentiates between categories.	The electron beam was transported <u>from</u> the cathode to the screen through a drift tube. A magnetic field separated the fast electrons <u>from</u> the slow electrons.
In	Inclusion in place, limits, time period or something abstract. Also used to indicate means.	The objective of this work was to identify the differences <u>in</u> dielectric properties between pure epoxy and epoxy/crepe paper composites. Forging was the accepted fabrication method <u>in</u> the early Iron Age. Soldering was commonly used <u>in</u> the electronics industry.

Preposition	Expresses	Example
Into	Transformation	The virgin surface was converted <u>into</u> a scratch resistant functional surface by ion implantation.
Like, such as	Similarity, examples	Polymers <u>like (such as)</u> PET and ABS are frequently used for packaging.
Of	Composition	The growth medium consisted <u>of</u> agar, sugar, and niacin.
On	On the subject of, concerning	Smith's work was <u>on</u> the effect of magnetic fields in hornet nesting.
Throughout	Temporal or spatial inclusion	The development of iron alloys proceeded at a slow pace <u>throughout</u> the Middle Ages.
To	Destination	The data sequence was sent <u>to</u> the processor via an optical fiber.
To	Purpose	A personal computer equipped with an A/D input card was used <u>to</u> record the data.
With	Inclusion	The sample <u>with</u> heated grains had the lowest nutritional value.
With	Means	The waveforms were observed <u>with</u> an oscilloscope

10.3.10 Commonly confused words

Table 10.34 explains the differences between words commonly confused in scientific writing and provides examples of their use.

Table 10.34. Commonly confused words.

Words	Definition	Examples
Accept	Reply positively or regard something as correct. (verb)	Five samples were <u>accepted</u> by the client for field testing. They all were Ti-based alloys, <u>except</u> for one which was Ni-based.
Except	Excluded, or left out. (preposition)	
Accurate	Errorless or exact (in scientific and engineering work, within some designated tolerance).	The measurements were <u>accurate</u> to within 0.1%. The accuracy was limited by the tertiary standard used for calibration. The measurements were <u>precise</u> within 0.03% and were limited by mechanical slippage. Five runs were conducted, and all results fell within the experimental error, demonstrating that the measurements are <u>reproducible</u>.
Precise	Conforms strictly to rule or proper form (in scientific and engineering work, repeatable within some designated tolerance).	
Reproducible	Can be repeated and obtain the same results. (all adjectives)	
Advise	To suggest or recommend (verb)	The scientific committee <u>advised</u> the ministry to adopt the proposed safety regulations, but the ministry did not heed this <u>advice</u>.
Advice	Suggestion or recommendation (noun)	
Affect	Influence or impact (verb)	The temperature <u>affected</u> the results due to the Sterling <u>effect</u>. Raising the temperature to 281 K <u>effected</u> a phase transition.
Effect	Result or condition resulting from (noun); Execute, do, cause, bring about (verb)	
Assay	A test to evaluate quality (noun)	The subject of Jane's senior <u>essay</u> was an <u>assay</u> of 52 Cu-bearing minerals found in the Timna vicinity.
Essay	A short piece of writing, a composition (noun)	
Attend	Listen closely, pay attention, participate in an event (verb)	His thesis advisor <u>intended</u> to <u>attend</u> the graduation ceremony.
Intend	Plan, be determined to do something (verb)	

Words	Definition	Examples
Compliment	Praise (verb)	The department head *complimented* the team leader on his choice of team members because they *complemented* each other's abilities so well.
Complement	Contribute different aspects to produce a more complete result (verb)	
Ensure	Make certain (verb)	The start-up company *insured* itself against product liability to *ensure* the financial stability of the venture.
Insure	Also means make certain but is more typically used with respect to insurance (paid policy against financial risk) (verb)	
Farther	Refers to greater physical distance	The rail gun could shoot the projectile *farther* with *further* charging.
Further	Refers to increased quantity or time	
Good	A common mistake is to use *good* — an adjective — to modify a verb: the correct adverb is *well*.	The electron gun performed *well*. *Good* performance and high throughput were obtained.
Well		
Imply	Suggest (verb)	The results *imply* that the frequency could be stabilized if the temperature could be further reduced.
Infer	Conclude (verb)	We *inferred* from the results that the frequency could be stabilized if the temperature could be further reduced.
Optimal	Most favorable (adjective, never used as a noun)	The *optimal* conditions for stable operation were bias voltage of -73.7 mV, and a load impedance of 13.4 Ω. This produced an *optimum* output power of 3.2 kW. At this *optimum*, a production rate of 3.4 kg/hr was sustained.
Optimum	Most favorable condition (noun, may be used also as an adjective)	
Principal	Person or thing of importance (noun) or main or dominant (adjective)	The college *principal* hired a new assistant whose *principal* task was to ensure compliance with the college's ethical *principles*.
Principle	A general truth, foundation of a system of beliefs, behavior, or chain of reasoning (always a noun)	

Words	Definition	Examples
Satisfying	Giving pleasure, satisfaction	*This novel control scheme has a <u>satisfactory</u> technical index with a transient process time of less than 0.28 s. The review committee found this result <u>satisfying</u> in light of the effort and resources expended in its attainment.*
Satisfactory	Meeting minimum requirements	
Than	Used to introduce the second element in a comparison (conjunction)	*First, those devices that performed slower <u>than</u> specified were eliminated. <u>Then</u> the system was re-tested.*
Then	Used to indicate time sequence — first, then (usually an adverb)	

10.3.11 *Abbreviations & acronyms*

An abbreviation consists of the first initials of a phrase such as HIV (human immunodeficiency virus). Each initial is pronounced separately. An acronym is usually formed by taking the first initials of a phrase or compound word and using those initials to form a word that acquires its own meaning and is pronounced as a word. Some familiar examples include AIDS (acquired immune deficiency syndrome), laser (light amplification by stimulated emission of radiation) and radar (radio detection and ranging).

Except for very common abbreviations or acronyms that are generally recognized internationally, spell out the full phrase at the first use and specify in parenthesis the short form you will be using throughout the paper. Thereafter, use the abbreviation or acronym consistently. Keep in mind that an international journal audience may not be familiar with all abbreviated terms that are in common use in your country. Although it is possible to search the internet for unfamiliar abbreviations or acronyms, often abbreviations and acronyms have multiple meanings; therefore, it is helpful to your reader to spell them out initially.

There are no hard and fast rules for using periods in either acronyms or abbreviations. It seems to be increasingly more common not to use periods. Within a paper, consistency is important. When an abbreviation that does have a period ends a sentence, no additional period is needed to end the sentence. Titles before names (e.g. Prof., Dr.) are abbreviated, with a period following. However, there is some debate about whether periods are used with titles used after names such as Ph.D., M.D., B.A., M.A., D.D.S. Follow the practice in your target journal. It is not customary to use titles before and after a name at the same time (i.e., *Dr. R.L. Boxman* or *R.L. Boxman, Ph.D.* but not *Dr. R.L. Boxman, PhD*). Do **not** abbreviate a title that isn't attached to a name: *We had a meeting with the prof.*

Abbreviations of common names of familiar institutions (e.g. MIT), countries (e.g. USA), corporations (e.g. IBM) famous people (e.g. FDR) and very familiar objects (e.g. TV, CD-ROM) are often used without explanation and without periods. Abbreviations of long, common phrases, such as IQ (Intelligence Quotient), rpm (revolutions per minute), mph (miles per hour), and mpg (miles per gallon) are generally acceptable even in formal academic texts and may be used without periods. However, units specified by the *Système international d'unités* (SI) are preferred, and their abbreviation and capitalization conventions should be followed.

Physical units used in technical writing have a space between the number and the units, e.g. *23 cm, 44 kg*. Keep the number on the same line as the unit by using a "sticky space" (control-shift-space in MS-WORD). When the measurement is used as a modifier, it is customary to put a hyphen between the number and the unit of measurement: a 15-m board, a 6-kg line, etc. Keep the number and its unit together on the same manuscript line by using the "sticky-hyphen" (control, shift, and — in MS-WORD). The SI units are not terminated with a period, nor is an *s* added to the abbreviation when the number is greater than one.

Abbreviations of Latin terms are commonly used in scientific writing. The most frequent are *etc.* (*et cetera* — and so forth), *i.e.* (*id est* — that is), *e.g.* (*exempli gratia* — for example), *et al.* (*et alia* — and others). Avoid overusing *etc.* – it leaves the reader to guess what else the authors had in mind. Instead be specific. The abbreviation *i.e.* (often translated as *in other words*) is often confused with *e.g.* The *i.e.* is used to introduce additional explanatory matter. If you can say *for example* as a substitute for the abbreviation, use *e.g.*, not *i.e.* Except for formal citations of previous research, it is better not to use *et al.* when you mean *and others*.

10.3.12 *Punctuation*

Punctuation aids correct interpretation. Complex sentences demand careful attention to punctuation. Simple, short sentence structures are easier to punctuate and are also usually more easily understood. A sentence ends with a period (.) or occasionally a question mark (?) but in scientific writing never in an exclamation mark (!).

A **comma** (,) is needed:
- following subordinate clauses or long introductory phrases before the main part of the sentence. *Because the signal to noise ratio was very low, the detector output was processed with a boxcar amplifier.*
- before and after subordinate clauses (nonrestrictive clauses) that are not essential to the meaning of the whole sentence. *The computer code, which was written in JAVA, was tested against five data sets.* A comma should never separate the subject and the verb of the sentence except for the two commas that set off an intervening subordinate or dependent clause. Similarly, dashes and parentheses may be used to set off information that is not an integral part of the sentence (i.e. the sentence can be understood without it).
- when a coordinating conjunction connects two independent clauses. The comma appears at the end of the first clause preceding the conjunction: *Most samples required 25 min of ultrasonic treatment, but the ALACXS sample only required 14 min.*
- between items of a serial list. Many writers will omit the comma before the final item prefaced by *and* unless the description of the items is lengthy. *The samples were washed, dried and annealed.*
- after transition words such as *however*.

A **semicolon** (;) may separate:
- items in complex series. *The samples were prepared by the following procedure: first, the substrates were cleaned with a detergent to remove grease and grime; then they were ground and polished using a sequence of abrasives ending in 0.1 μm diamond powder; then the substrates were ultrasonically cleaned in methanol and mounted in the vacuum deposition chamber; multi-layer TiN/NbN coatings were*

sputter deposited onto the substrates; and finally the coating samples were annealed in vacuum at 960 K for 12 minutes.
- sentences that are closely related but are not linked by a coordinating conjunction.

A **colon** (:) may precede:
- a detailed list.
- a quote.

Double quotation marks may be placed around direct quotes from another writer or speaker. Single quotation marks or brackets are used for a quote within a quote. The source must be cited. However, direct quotations are rare in the physical sciences.

Numbered and bulleted lists are useful to distinctly display lists (e.g. of materials, characteristics, actions, steps taken, main conclusions).
- Each element in the list is emphasized by preceding it by a bullet point (e.g. •), a number (e.g. *1*) or a letter (e.g. *a*). Bulleted items are displayed on separate text lines; numbered items can be either on separate text lines, or within a paragraph. A numbered list within a paragraph is usually preferred because it efficiently utilizes page space.
- Use a numbered list when there is some importance to the total number or order of the items or if it will be convenient later in the text to refer to the items by number. Bulleted lists are common on lecture slides, but rare in research reports. They can, however, be used effectively to emphasize main points, e.g. the main conclusions.
- Bulleted or numbered items displayed on separate lines should be punctuated with a period if each item is a complete sentence or left unpunctuated if not.
- Numbered items within a paragraph should be separated by commas if they are single words or short phrases. If they are long or complex phrases, they may be separated with semi-colons. If each item is a complete sentence or several sentences, end each sentence with a period and begin each sentence with a capital letter.

10.3.13 *Capitalization*

Capitalized words in English include:
- The first word of a sentence.
- Proper nouns — names of specific people, places, and institutions — are always capitalized, no matter where they occur in the sentence (e.g. *Raymond, Dr. Boxman, Boston, Boston University, Fifth Avenue, New England, the United States, the Mauna Kea Observatories*).
- The main words in titles of books, articles, research papers, chapters and sections, as well as songs, plays and works of art, are capitalized. Articles, prepositions and conjunctions are rarely capitalized if they are not the first word of the title. However, it is becoming common for only the first word to be capitalized in the title of scientific journal articles.
- The first word in bullet lists may be capitalized when the items in the list are lengthy or complex, even if they are not complete sentences.
- The pronoun *I* is always capitalized.
- When an adjective owes its origins to a proper noun, it is normally capitalized. Thus we write about *Cartesian coordinates, Ohmic heating, Galilean relativity, Portland cement*.

Common nouns are words for a general class of people, places, things, and ideas (*engineer, laboratory, telescope, diffusion*). They are only capitalized if they are the first word in the sentence or if they are part of a title. They are never capitalized to emphasize their importance within the sentence.

If your native language does not normally use capitalization as is typical in many Middle Eastern and Asian languages, you may need to pay special attention to capitalization.

10.3.14 *Mathematical grammar*

Scientific writing frequently uses mathematical expressions, both within lines of text, or on separate lines (and in this case, they are often numbered). In both cases, the equations can be treated as integral parts of sentences, wherein operators such as $+$, $-$, $\times$ act as prepositions or verbs, other operators such as ∏, ∑, ∇, act as nouns, and equality and inequality symbols (=, ≤, >, etc.) act as verbs. This allows the mathematical expressions to be vocalized with words. Sentences containing mathematical expressions should be punctuated as if the vocalization was written. Some examples are given in Table 10.37.

Table 10.37 Mathematical expressions

Text with math expression	Vocalized text	Comments
If $x>y$, then $z=A+b\times c$.	If x is greater than y, then z equals A plus b multiplied by c.	Note the use of the comma that separates the condition from the consequence, both in the mathematical and vocalized examples.
The force density on the fluid is given by: $$f=f_v+f_e+f_m \quad (3)$$	The force density on the fluid is given by: f equals f_v plus f_e plus f_m.	Note the period after the equation. In placing punctuation marks, the equation number is ignored.
The force density on the fluid is given by: $$f=f_v+f_e+f_m, \quad (3)$$ where f_v is the viscous force density, f_e is the electrical force density, and f_m is the magnetic force density.	The force density on the fluid is given by: f is equal to f_v plus f_e plus f_m, where f_v is the viscous force density, f_e is the electrical force density, and f_m is the magnetic force density.	A comma follows the equation, and <u>where</u> is not capitalized, as it is the continuation of the sentence.

10.4 Especially (but not just) for non-native speakers

The tips given here are oriented at the speaker of English as a second language (ESL) but they may also be useful for native speakers of English. In teaching scientific writing around the globe, we have seen that both native and non-native English speakers face many of the same challenges when writing up research. Native English speakers typically have an intuitive understanding of correct usage, but they do not necessarily have a sound grounding in the formal rules of grammar. In contrast, ESL students must rely more on formal rules, and often they learned grammar with more structure than native English speakers. Good writing, and in particular, clear scientific writing, is not guaranteed by writing in your mother tongue.

English in technical papers is easy compared to that in general literature. Technical papers use a limited vocabulary and limited grammatical forms. Moreover, technical English is highly conventional. Clarity is more important than literary style. Thus learning a relatively small subset of skills, grammar and vocabulary can greatly improve your ability to generate clearly understandable research reports.

10.4.1 *Learning by imitation*

Becoming well acquainted with the literature in your field is an essential part of your education and research. It is also an opportunity to adopt the best scientific English in your own work. As you read, choose several good papers as examples. These papers should be the same type of paper as you will be writing — experimental or theoretical — and in your field or closely related to it. Look for papers that are clearly written and enjoyable to read. The author should preferably be a native English speaker. Use these articles as templates for your own writing.

Take note of key sentences in each section of the research report. It can be helpful to make a list of commonly used vocabulary specific to your field. Try to imitate the style. Write sentences by copying their structure, and substituting words appropriate to your research. As long as the content is your own, this is not plagiarism.

10.4.2 *Dictionary use*

A good English dictionary can be helpful in clarifying the range of meanings denoted by a word. However, use a dual-language dictionary with extreme caution. This also applies to the language tools (e.g. dictionaries and thesaurus) built into your word processor. It is safest to only use words that you have seen in a similar context in another paper in your field. Nuances often do not translate well from language to language. Some terms used in other languages are adaptations of English technical term. This adaptation may not appear in the dictionary. And even when

> **English**
>
> English has become the *de-facto* international language of science, technology, and business, but it is not a simple language. English has more than 500,000 words. One of the complicating factors is that English words generally stem from two systems of roots: Germanic and Latin. The Germanic words were introduced by Germanic tribes that populated the British Isles. Typically, common household and agricultural words have Germanic roots and hence their similarity to modern German words (e.g. *house-haus, home-heim, oven-ofen, mouse-maus, cow-kuh, plough-pflug*). Words with Latin roots were introduced from Old French during the more recent Norman conquest of England. Typically, more sophisticated words have Latin roots. Often English will have two words expressing approximately the same thing, one from a Germanic root and the other from a Latin root, e.g. *home* and *domicile*. Fortunately, only a small vocabulary and a subset of grammatical conventions is needed for most scientific writing.
>
> English has significantly evolved over time. A modern English speaker will have some difficulty understanding the Elizabethan English of Shakespeare (1564–1616), considerable difficulty in understanding the Middle English of Chaucer (~1343–1400), and overwhelming difficulty in reading the Old English epic poem Beowulf (written sometime between 700–1025).

the words in both languages are similar, they may not have exactly the same connotation. Often, one word may have several meanings in one language and in another language, multiple words each express one of the meanings. For example:

In Chinese, 有效的 has multiple meanings which are expressed in English as *effective, efficient or valid*.

In Danish, *stamme* has multiple meanings which are expressed in English as *stem, log, originate, tribe or stutter*.

In English, *fan* may mean an enthusiastic supporter of a sports team or a mechanical device providing a flow of air. In Hebrew, the sport fan is termed an אוהד (*ohaid*) and the mechanical device a מניפה (*m'nifah*).

In German, *spannung* has multiple meanings that are expressed in English as *voltage difference, potential difference, (residual) stress* or *tension*.

In Hebrew, מנה (*mana*) has multiple meanings that are expressed in English as portion, ration or counted.

In Russian, Возмущение (*vozmushcheniye*) has multiple meanings that are expressed in English as a perturbation in mathematics and theoretical physics, and may be expressed in everyday language as disturbance, indignation or outcry.

10.4.3 *Keeping it simple*

Remember some basic guidelines:
- Decide what you want to express and in what order.
- Work from a detailed outline — down to the level of the paragraph.
- Keep your sentences simple and short, and use the natural English word order: subject, verb, predicate. Avoid compound sentences and complex constructions.
- Use the most basic verb tenses — most sentences should be in the simple past or present.
- Choose simple words. Avoid redundant phrases. Many writers have the misconception that more complex sentence constructions and vocabulary sound better. This is only true in scientific writing if complexity contributes to clarity.

KISS

Keep it Stupid Simple, **KISS,** is an organizational principle that works well in many aspects of life, including scientific writing.

Regional Differences in English

English varies slightly depending on country and regions within a country. English spread around the world with the expansion of the British Empire (as did Greek and Latin with earlier empires). Some of the former colonies have primarily English speaking populations (e.g. U.S., Canada, Australia, New Zealand) and English is retained as either an official or administrative language in other former colonies having an ethnic and language mixture (e.g. India, South Africa, Singapore).

The biggest difference is in pronunciation. The differences are sufficiently great that two English speakers from different regions may have difficulty in understanding each other's speech.

There are also minor differences in vocabulary and spelling. Some examples are given in Table 10.36 below. These differences should not greatly affect scientific writing, except that it is good form to adapt either U.K. or U.S. vocabulary and spelling consistently within a given paper.

Table 10.36. Examples of differences between U.S. and U.K. vocabulary and spelling.

Vocabulary		Spelling	
U.S.	U.K	U.S.	U.K
aluminum	aluminium	catalog	catalogue
antenna, aerial	aerial	color	colour
battery	accumulator	defense	defence
counterclockwise	anticlockwise	gray	grey
faucet	tap	imbed	embed
flashlight	torch	ionize	ionise
ground fault circuit interrupter	residual-current device	analyze	analyse
ground wire	earth wire	meter	metre
jackhammer	pneumatic drill	fiber	fibre
kerosene	paraffin	program	programme
line	mains	signaling	signalling
lumber	timber	sulfur	sulphur
mass transit	public transport		
math	maths		
mononucleosis	glandular fever		
plexiglas	perspex		
tube	valve		
high-voltage transmission line towers	pylon		
wrench	spanner		

10.4.4 *Finding and working with a mentor*

If your institution has a writing center or any other organized system for tutoring writing, take full advantage of it. Many universities offer special courses to ESL speakers and/or formal instruction in writing up research. Some provide writing tutors with expertise in various fields. The time to use such facilities is not the month before your thesis is due, but early on and throughout your studies.

Ideally, your thesis advisor will be a key mentor. He or she will review drafts of your thesis, as well as any articles you submit under joint authorship. Sometimes, another senior member of the group or laboratory may be a willing and helpful writing mentor. Often such individuals are in demand as joint authors.

In all cases, the benefit you will get from your mentors will be directly proportional to the effort you yourself put into your writing. Generally, language problems are relatively easy to correct. However, correcting scientific thinking and report organization are difficult. Therefore, strive to get the thinking and organization right, before asking a writing mentor to look at a manuscript. Discuss your scientific thinking, the scientific content and the organization of a proposed report with your scientific mentor, and agree on an outline before you start to write.

If you are reading this text, most probably writing in English will be an important and continuing activity throughout your career. Accordingly, if you have the opportunity to work with a mentor, use the opportunity to learn how to write better. Ask your mentor to read one section of your report and to explain the corrections. Then apply the same corrections to the next section, before you submit it to your mentor, and see if you understood the principle involved. When you are in a hurry to finish a short-term written work, it may be very tempting to use a mentor merely to correct your writing. However, most mentors are far more willing to help the student who demonstrates an ongoing effort to learn from them. Moreover, first understanding and only then applying the corrections will improve your writing for years to come.

For further reading:

Robert E. Berger, *A Scientific Approach to Writing for Engineers and Scientists*, IEEE Press/Wiley, 2014.

The Chicago Manual of Style, 15th edition, University of Chicago Press, 2003.

Robert Hartwell Fiske, *To the Point: A Dictionary of Concise Writing*, W.W. Norton & Company, 2014

William Strunk Jr. and Stanford Pritchard, *The Elements of Style, Annotated and Updated for Present Day Use*, Second edition, Springside Books, 2012.

Index

Abbreviations, 35, 68, 157, 252, 253
Abstract,
    in business plan, executive
        summary, 126
    in patent application,
        144–145
    in poster, 97, 99, 101
    in research proposal, 108
    in research report, 13, 25,
        53–54, 66–68, 70, 75
Acceptance (of paper), 77, 79
Acknowledgement, 69
Acronyms, 68, 93, 102, 252
Adjectives, 33, 197, 201, 209, 227, 228,
    234–237, 245, 246, 250, 251, 256
Adjudicator, 76
Adverb, 201, 237, 238, 251, 252
American vs British, spelling
    and terms, 261
Animation, 89, 92, 93
Answer to the Research Question, 7–9,
    13, 26, 30–32, 39, 54, 54, 65, 87,
    98

Apparatus
    in patent application, 148
    in posters, 98
    in research proposal, 117
    in research report, 11, 12,
        32–37, 56, 60
        special, custom, 34-37
        standard, well-know,
            34
    writing conventions,
        186, 205, 213–217, 222,
        234
Appendix, 58, 134, 135
Archimedes, 1–3, 83
Articles, part of speech, 227, 232–234,
    236, 256
Attire, 101
Attitude, author's, 21, 22, 210
Audience, 16, 85–88, 89, 92–97, 103,
    109, 145, 154, 160, 161, 252
Auditorium, 87, 88, 92, 94, 96, 102, 103
Author, who is and isn't, 68–70
Axis, 43, 45

Background
    in business proposal, 129, 130
    in Introduction of research report, 11–13, 16–18, 39, 67, 68, 70
    in lecture, 93
    in patent application, 147
    in poster, 99
    in research proposal, 107–110, 114, 122
    introductory sentence, 16, 17
Backup, 187, 188
Balance Sheet, 133
Blind man's rule, 47, 56
Body,
    in correspondence, 75, 77, 82, 163, 165–168
    in business plan, 134
    in research proposal, 108
    in research report, 12, 13, 53–55, 58, 71
Body text, 35, 37, 39, 42–47, 50, 90, 189, 191, 198
Boss, 68, 180
Budget, 115–119, 123, 132
Bullet list, bullet points, 90, 97–99, 168, 194, 255, 256
Burn rate, 132, 134
Business cards, 100, 101
Business letter, 75, 82, 163–168, 181
Business plan, 124–136, 214

Capitalization, 253, 256
Caption, 35, 37, 38, 40–42, 44–47, 190, 191, 198

Career, 101, 169, 170, 179, 182, 183, 186, 262
Cash Flow, 133
Citation
    automatic numbering, 190
    in lectures, 86
    in research proposal, 113
    in patent application, 145
    management, 22, 190
    no plagiarism, 73
    of author's own work, 24
    styles, 22
    symbols – don't use as words, 21
    types
        author prominent, 20
        information prominent, 19, 20
        state of the art, 21
        weak author prominent, 20
Claim,
    in patent, 138, 144, 148–153
    in research report, 9, 159
    writing conventions, verbs supporting or rejecting, 225, 226
Clause
    independent, main, 61, 198, 199, 200, 254
    restrictive vs non-restrictive, 243, 254
    subordinate, 200, 254
Clips, 89, 92, 95, 98, 161
Collaborator, 100, 187, 188, 191
Comment
    comment sentence, 44–46, 49–53, 56, 57
    reviewers' and editor's, 69, 76–82

writing conventions, 215
Common problems, 10, 40, 58, 119, 120, 135
Company, 86, 127–135, 139, 144, 145, 157, 180, 181
Comparative, 235–237
Comparison, 41–44, 49, 60–62, 92, 130, 157, 215, 225, 235, 240, 245, 252
Competition, 106, 128, 151
Composition, 192–208
Compound sentence, 206, 238, 239, 260
Conclusions,
    in business plan, 134
    in lecture, 86, 87
    in patents, 150
    in popular media, 156
    in poster, 97, 99
    in research report, 7, 9, 11, 13, 30–32, 54, 55, 65–67, 70, 71
    in slides, 89, 92
    word choice, 214, 224, 225, 242
Conference,
    abstract, 68, 97, 99, 101
    auditorium, 103
    briefing chairs and speakers, 102, 103
    conduct, behavior, 100, 101
    food service, 104
    inclusion in CV, 170
    lecture (see also lecture), 85–96, 198, 254
    planning, 101–104
    poster, 96–99, 103, 104
    speaker, 86, 94–96, 102, 103
Conjunctions
    coordinating, 236, 237, 253, 254

    correlative, 237, 238
    subordinating, 240–243, 252
Consistency,
    in business plan, 135, 136
    in research proposal, 123
    in research report, 30–32, 35, 60, 192, 252
Consistent terminology, 207, 208
Contacts, 84, 100, 130, 180
Copy editor, 79
Copyright, 73, 74, 76, 79, 139, 140
Cover letter, 75, 77, 80, 82, 83, 166, 167
Cures, 96, 120
Cursor, 94–96, 103
Customers, 124, 126–131, 134, 135, 138, 180
Curricula Vitae (CV), 107, 118, 134, 161, 168–171, 175–177, 181

Data, 38, 41–43, 60, 72, 134, 230, 238
Degree of certainty, see also explanations, scale of certainty, 62, 63, 224, 225
Demonstratives, 230, 232
Dependent claim, 150–152
Derivation, of equations (see also Equation), 12, 32, 38, 54, 55, 58, 92, 122, 157
Detail, required level, 6, 10, 12, 33, 34, 35, 38, 39, 40, 48, 55, 56, 58, 70, 85–87, 92, 98, 112, 113, 115, 118, 121, 122, 131–135, 138, 140, 145, 149, 151, 153, 158, 159
Determiner, 232, 233, 245, 246
Diagram, 33, 35, 36, 89, 92, 149, 208

Dictionary, 258
Direct object, 196, 197, 227–231
Disclaimer, 149
Discussion,
    in conference, 97–102
    in lecture, 86, 87
    in poster, 97, 98
    in research report, 11, 13–15, 39, 48, 49, 50, 53, 54, 59–65, 70, 71
    in slides, 88
    word choice, 210, 213–216, 224, 225, 235
Discussion period, 100
Distraction, 185
Document naming and tracking, 187
Double publication, 73, 74
Drama, 155, 156
Drawings, 89, 92, 95, 144, 145, 148, 149, 151, 152
Duplicate, 12, 15, 33, 55, 87, 140

Editor, 74–82, 155
Education, 168, 169, 258
Electronic correspondence, 167, 168
Embodiment, 146, 149, 150
Employer, 100, 137, 144, 169, 179–181, 184
English (language), 209, 259, 261
English as a second language/ESL, 58, 192, 196–198, 209, 232, 236, 243, 246, 257–262
Equation, 12, 32, 38, 54-56, 58, 89, 92, 98, 122, 157, 190, 191, 214, 256, 257
Equipment, 33-37, 69, 70, 102, 103, 108, 114–118, 123, 130, 132, 133, 178, 217, 230, 232

Ethical issues, 72–74
Evaluation, 68, 76, 80–82, 118–120, 124
Executive summary, 126
Expected result, 111–112
Experimental apparatus/set-up/method/procedure — see Methodology
Experience, 107, 125, 128, 166, 168, 169, 173, 176, 178, 179, 181
Explanation, 15, 37, 38, 48, 49, 50, 54, 55, 58, 60–64, 78, 87, 88, 134, 135, 149, 159, 160, 190, 214, 215
Eye contact, 94, 96

Facilities, 70, 106, 108, 114, 117, 131
Figure, 9, 35–37, 39, 40–48, 50–53, 56, 57, 73, 79, 89–91, 96, 190, 214, 217
    blind man's rule, 47, 56
    heads-up display, 35, 40
    illiterate man's rule, 40, 56
Financial plan, 132–134, 135
Fine tuning, 85, 93
Functional description, 35, 36
Funders, 124, 125
Future research plans/directions/work, 62, 71, 112, 122, 156, 178, 210–212, 213, 215, 218, 227

Galley proofs, 74, 76, 79
Gantt, 113
Gap, 11, 13, 18, 22–26, 29–32, 64, 71, 109, 110, 120, 129, 147, 156
Generalization, 49, 60, 63, 215
Gerund, 228-229

Government/Governmental, 106, 124–126, 135, 137, 138, 145, 154
Grantor, 106–108
Graph, 40–45, 91, 113
Graphics, 85, 87–93, 96, 98, 190
Guidelines,
    communicating with media and general public, 156–157
    figures and tables, 40–41
    graphs, 42–43
    keeping it simple (KISS), 260
    lecture presentation – common problems and cures, 96
    modesty in scientific writing, 64
    poster sessions, 104
    reader friendliness, 10–12
    research proposals – common problems and cures, 120–123
    reviewing a paper, 80–83
    sentence composition, 195–208
    slide preparation, 88–89
    standard apparatus, identification and description, 34
    tables, 41

Heads-up display, 35, 40, 56, 91
Hierarchical structure, 192, 193, 195
Humor, 87
Hypothesis, 9, 54, 58, 60, 66, 213

Illiterate man's rule, 40, 56
Implications, 14, 22, 60, 63, 65, 87, 92, 158, 215
Indicative sentence, 66, 67
Indirect object, 196, 197, 227
Infinitive, 209, 228–230

Inflection, 96
Information, 10, 12, 15, 22, 35, 37, 40, 49, 59, 65, 95, 101, 102, 104, 108, 138, 139, 156, 201, 234, 243, 245, 254
    old vs new information, 18, 193, 197, 198, 217, 230, 231, 233
    information vs knowledge, 15, 60
Informative sentence, 21, 67
Integrity, 72, 73
Intellectual property (IP), 73, 126, 127, 128, 133, 139
Internal report, 6, 7, 12, 16, 19
Interpretation, 15, 48, 50, 60, 68, 254
Interview, 152–153, 155, 158–161, 180–183
Introduction,
    in business plan, 129, 130
    in lecture, 86, 87
    in patent application, 147, 148
    in research proposal, 109–111
    in research report, 7, 11, 13, 16–32, 39, 53, 54, 64, 67, 71, 214, 218
Introductory phrase, 200, 201, 254
Introductory sentence, 16, 17
Invention, 137–153, 155
Inventor, 137–151, 171
Investment, 108, 132, 134
Investor, 124–129, 132–136, 139

Jargon, 109, 120, 146, 157, 158
Job hunting, 168, 179–184
Journal choice, 83

Journal paper, 6–71, 72–83, 139, 144, 145, 170, 186
Justification,
    of budget, 116–118, 123
    of method or result, 60, 158
    of paper, 38–39

Keep with next, 189
Key personnel,
    business plans, 126, 127, 134
    job search, 180
    research proposals, 114, 115, 118
Key words, 63, 68, 89, 92, 98, 193
Knowledge, 15, 22, 25, 59, 86, 107, 130, 138, 151, 234

Laser pointer, 94, 96
Leadership, 169, 170, 183
Lecture
    delivery, 94–96
    preparation, 85–93
    slides, 87–93, 96, 198, 255
Letterhead, 75, 82, 163–165
Limitation, 60, 70, 150, 152, 158
Literature review,
    in lecture, 86
    in popular media, 156
    in research proposals, 109–110, 120
    in research report, 10, 13, 18–23, 73, 81
    word choice, 210, 211, 214
    writing discipline, 186
Location (placement considerations), 59, 77, 78, 82, 103, 104, 109, 244, 246

Location sentence, 40, 44-46, 47, 49, 50–53, 56, 214, 217

Magazine, 155, 157-158
Main thought, 199
Market, 125, 126, 129, 130, 131, 133, 135
Market survey, 109, 130, 134, 214
Mathematical grammar, 256-257
Mathematical proof, 53, 54, 58
Means, vs objective, 27, 110, 111, 201, 202, 248, 249
Media, 68, 154–156, 159, 160, 171
Mentor, 262
Metaphor, 156
Methodology, methods,
    analogy in business plan, 131
    analogy in patent applications, 149
    in abstract, 67, 68
    in conference presentations, 86, 87, 92, 93, 99
    in research proposals, 107, 108, 112–114, 121–123
    in research report, 12, 13, 15, 27, 32–40, 54–56, 60, 66, 70, 71
    word choice, 205, 210, 213, 220–222, 234
Methods and Materials – see Methodology
Metrics, 83
Micrograph, 45, 46, 92
Microphone, 95, 96, 102, 103
Milestone, 132
Modal auxiliary, 22, 29, 62, 212, 227

Model assumptions, 32, 54, 55
Modesty, 29, 64, 216
Monopoly, 138, 139
Multi-media, 83, 95

National phase, 143
Natural verb form, 198, 199
Natural [English] word order, 196–198
Networking, 100, 186
New information, 18, 197
Newspaper, 155, 158, 159
Nomenclature, 35, 58-59, 188
No-show, 101, 102
Not obvious (patent criterion), 141, 142, 148, 153
Noun, 195, 196, 198, 199, 218, 227–230, 232, 233, 236, 237, 245, 257
    compound, 228, 236, 246
    countable, uncountable, 230, 232, 236
    derived from verb, 198, 228-230
    number, 218, 228
    proper vs common, 253, 256
Noun phrase, 195, 227, 230, 232
Novelty, 23, 29, 56, 81, 107, 109, 110, 113, 140, 141, 143, 147, 150, 151, 152, 153, 157
Numbered list, 255
Numerical data or values, 41-43, 49

Object, 197, 210, 215, 228, 229, 245

Objective (see also Statement of Purpose)
    in conference presentations, 86, 99
    in job search documents, 169, 170, 177
    in popular media, 158
    in research reports, 11, 26–32, 58, 60, 66, 67, 70, 71
    in research proposals, 108, 110–111, 114, 118, 121–123
Objective, vs. means, 27, 111
Objectivity/Objective (adjective), 15, 64, 71, 157
Onion strategy, 150
Open access, 74, 79, 83
Operational plan, 131, 135
Outline, 10, 11, 68, 86, 186, 191, 193, 260, 262
Overhead, 116, 117, 118, 133

Page charges, 74, 79
Paragraph, composition/structure, 10, 189-191, 192–195, 207, 208, 260
Parameters, 38, 39, 40, 42, 43, 45, 47, 48, 56, 147, 149, 150
Participle, 21, 29, 213, 228, 237
Patent, 37, 73, 109, 126, 128, 137–153, 171
    jargon, 146
    patentability, 140-142, 143

patent application, 137-153
PCT, 143
Peer review, 74, 80
Person, 64, 216, 218
Photograph, 35, 45, 46, 87–89, 92, 95, 161, 190
Physical model, 53–57
Plagiarism, 73, 258
Plan
    business, 124–136, 214
        financial, in business plan, 132–134
        operational, in business plan, 131
    research, in job search, 178
    work, in research proposal, 107, 108, 112–114, 118, 122, 123, 210, 211, 213, 215
Possessive, 227, 228, 230, 231, 232
Poster, 84, 96–99, 101, 103, 104, 198
PowerPoint, 92, 94, 97
Practitioner, 12, 84, 86, 111
Predicate, 195–197, 204, 209, 260
Preposition, 229, 245-249, 256
Prepositional phrase, 196, 197, 201, 227, 229, 237
Presentation (see also Lecture)
    presentation sentence, 47–53, 56, 214, 217
    presentation standards, 35, 40, 43, 81
Press release, 155, 158, 159
Preview, 13, 29, 30, 86, 111, 148
Prior art, 142, 145–147, 150, 151
Probability of success, 107, 108, 119, 125

Procedure, 10, 32, 33, 38–39, 45, 48, 59, 139, 140, 186, 213, 214, 217, 220, 222
Product, 64, 84, 86, 107, 121, 125, 126, 128–135, 138, 139, 140, 154, 157, 180
Profit, 125, 126, 129, 132–134, 136, 138, 139, 153
Profit and loss statement, 133
Program,
    computer applications, 33
    citation manager, 22
    patentability, 140
    slide preparation, 88, 89, 93, 97
    word processor, 168, 188
    conference, 84, 86, 98, 101–104
    granting agency, 106, 107, 119
    research, 108, 118, 213
    work, 121, 122
Pronouns, 195, 209, 227, 230-232, 236, 243, 245, 256
Proof see also mathematical proof, 58, 62, 63
Proposal, 106–125, 131, 210, 211, 213
Protocol, 38, 101
Provisional patent application, 142, 143, 151
Public disclosure, records, 138, 141, 145, 151
Public (general public), 84, 155, 156–162
Public funding agency, institution, or sector, 106, 119, 154
Public policy, 65

Public relations, 154
Publication process, 74–80
Publicity, 102, 154
Punctuation, 254-255, 257

Qualifiers, 201, 203

Radio, 159, 160
Reader,
    analogy to radio receiver, 2
    reader friendliness, meeting reader expectations and assumptions, 3, 10, 12, 13, 15, 16, 18, 21, 22, 29, 30, 37–43, 45, 47, 50, 54, 58, 59, 60, 62, 64, 65, 67, 68, 78, 109, 125, 135, 148, 149, 152, 167, 168, 193, 196, 200, 201, 203, 207–209, 230–235, 238, 252, 253
    who are your readers? 12, 125, 145, 156, 153, 159, 162
Rebuttal, 77, 78, 152
Recommendations, 60, 63, 64, 74, 76, 80, 81, 182
Reduced to practice, 140, 141
Referee, 18, 23, 78, 106, 119–121, 123
References, 17, 19–22, 25, 40, 42, 44, 47, 55, 56, 59–61, 66–68, 74, 113, 114, 163, 164, 170, 190, 191, 214
Regulatory, 126, 129–131
Rehearsal, 85, 93, 96, 183
Rejection, 23, 62, 76, 78

Reminders, 60
Reporter, 155, 158, 159
Research proposal, see also Proposal, 106–123, 124–127, 129, 131, 132, 210, 211, 213, 215
Research question, 7–9, 13, 26–28, 30–32, 39, 54, 60, 64–66, 87, 98, 214
Research report, 6–73, 86, 108–113, 137, 144, 145, 147–150, 153, 155, 158, 187–189, 192, 210, 215, 218, 255, 258
Research statement, 178
Research story, 157
Resources, 106–108, 114, 123, 126, 127, 131, 132, 178
Resubmit, 78
Results,
    in lecture, 84, 86, 87, 92, 93
    in popular media, 156–159
    in research report, 6, 7, 9, 11–15, 21, 33, 34, 37–51, 55–57, 59–62, 64, 65, 67, 69, 70, 72, 73, 81
    preliminary, in research proposal, 109, 111, 113, 185
    word choice, 192, 201, 203, 205, 210, 211, 213–215, 218, 223, 224, 234, 235
Résumé, 5, 127, 134, 168–172, 177, 180, 181
Return on investment, 125, 135
Review paper, 70, 71
Review process, 72–83, 163
Reviewer, 23, 33, 69, 74–82, 118
Reviewing a paper, 80–83

Revision, 10, 76–82, 187, 191

Salutation, 163, 165, 166, 168
Scale, in photographs and micrographs, 43, 45, 46, 89
Scale of certainty, see also Explanations and Degree of certainty, 62, 63
Schedule, 25, 85, 86, 98,185
Schematic diagram, 35
Secrecy, 138
Seminar, 84, 85, 93, 183, 184, 186
Sentence connectors, 241–243
Sentence structure, 67, 96
    completeness, 195, 196, 198
    compound sentence, 206, 238, 239, 260
    cures for sick sentences, 195–208
    main thought, emphasize, 199–201
    natural verb form, 198, 199
    natural word order, 196, 197
    old before new information, 196, 217
    parallel structure, 206, 207
    phrases to omit or shorten, 201–206
    precision, 205, 206
    predicate, 195, 196, 204, 209, 260
    subject, 19–21, 151, 195–199, 201, 204–206, 208–210, 215, 216, 218, 227, 228, 230, 231, 254, 260

Sentence types,
    citations, in literature review, 18–21
    comment, in Results, 44, 53, 56, 57
    gap, in Introduction, 22–26, 31, 64, 109–110
    indicative, in Abstract, Conclusions, 66–68
    informative, in Abstract, Conclusions, 21, 31, 65, 67, 68
    location, in Results, 40, 44–53, 56, 57, 214, 217
    opening, in Introduction, 16–18, 110, 120
    presentation, in Results, 44–53, 56, 57, 217
    preview, in Introduction, 29, 30
    Statement of Purpose, Objective, in Introduction, 26–28, 31, 110, 211, 212
    value statement, in Introduction, 28, 29,31
Services, 70, 84, 86, 116, 117, 121, 125, 126, 128–130, 132, 133, 135, 158, 180
Signature, 163, 165–168
Significance, 28, 39, 59, 60, 64, 72, 81, 108, 111, 114, 121, 122, 158, 159
Significant figures, 41, 42
Slide, 87–97, 103, 168, 255
Social contract, 137
Solution method, 54–56

Speaker,
    non-native, of English, 4, 5, 192, 196, 198, 199, 228, 232, 236, 246, 257–262
    at a conference, see also Lecture, 86, 94–96, 102, 103
Specification, 33, 144–146, 149–151, 153
State of the art, 21, 22, 109
Statement of Purpose, 7, 13, 18, 23, 26–31, 110, 148, 213, 214
Sticky-hyphen, 190, 253
Sticky-space, 190, 253
Subject,
    area, of proposed research, 25, 109, 114, 115, 119, 120, 125
    area, of conference or meeting, 102
    of paper or document, 25, 68, 163–165, 167, 168
    of sentence, 19–21, 151, 195–199, 201, 204–206, 208–210, 215, 216, 218, 227, 228, 230, 231, 254, 260
Submission, 69, 75, 76, 80, 102, 152, 163, 166
Superlative, 181, 235–237
Symbols, 38–40, 55, 58, 59, 91, 92, 257

Table, 9, 38–43, 47, 48, 56, 53, 59, 73, 77–79, 130, 133, 134, 190, 191, 214
Teaching statement, 178

Team, 6, 102, 107, 108, 118, 123, 125, 128
Television, 159–161
Tenure track, 178, 179, 182
Theoretical paper, 7, 32, 53–59, 62, 92
Thesis, 12, 16, 19, 66, 106, 188, 262
Timing in oral presentations, 93
Title,
    of report or document, or section thereof, 22, 25, 32, 33, 40, 47, 68, 70, 75, 88–92, 97, 99–100, 112, 144, 170, 171, 186–188, 256
    personal, 164–169, 253
Topic, 3, 16, 17, 23, 25, 64, 70, 71, 84, 97, 98, 104, 107, 109, 112, 119, 120, 152, 158, 160, 161, 193–195, 207, 208
Trade journal, 109, 155, 157
Trade secrets, 128, 139
Trademarks, 73, 128, 139
Travel, 115, 117
Typesetting, 74, 79

Unexpected findings, 73, 157
Units, 41, 43–45, 190, 253
Utility, 142, 143

Validation, 60, 61
Value statement, 13, 28–32, 111, 128, 215
Variable, 38–40, 42–43
Verbs
    action, 195, 196, 209, 220
    authors' attitude, 21, 62, 210

    commonly used, 218–227
    generalized, 198
    in mathematical expression, 256, 257
    in patent claims, 151
    in sentence, 195–198, 228, 229, 254, 260
    modal auxiliary, 29, 49, 62, 209
    natural form, 198, 199
    number, 209, 210, 218, 230
    stative, 195, 196, 209
    tense, 21, 49, 61, 62, 209–215
    voice, 209, 210, 215–217
Version, 25, 135, 187–189
Visual aids, 48, 87, 161, 193
Voice volume, 95, 96

Website (also web, web pages, online mechanism), 22, 25, 75, 80, 82, 83, 100, 127, 130, 131, 155, 161, 164, 168, 171, 180
Web of Science, 25
Word choice, 192
    commonly confused words, 246–249
    natural verb form, 198, 199
    parallel structure, 206
    precise terms, 205, 206
    unnecessary words, 201–205, 208
Word processor tools, 188–192, 258
Work plan, 107, 108, 112–114, 118, 122, 123, 131, 210, 211, 213, 215
Writing discipline, 185–187